AF811007

Marine management in disputed areas

As world natural resources diminish and the necessity of protecting our environment becomes increasingly urgent, the need for efficient marine management is more essential now than ever before. However, marine boundaries are not easily defined, and in disputed areas the prospect of sound management becomes ever more difficult.

To illustrate these difficulties Robin Churchill and Geir Ulfstein have chosen the case of the Barents Sea – a near perfect example. This is an area rich in living resources and is widely believed to contain considerable quantities of oil and gas. Despite this wealth, however, the living resources are over-exploited and the area's ecosystems are under growing threat. This lack of efficient management can partly be attributed to general difficulties in international resource management. Marine management in this area is, however, complicated by two legal disputes. The first is between the two riparian states of the Barents – the Soviet Union and Norway – over where their respective maritime boundaries should lie. The second one is between Norway and most of the 40 or so states party to the 1920 Spitsbergen Treaty. Under this Treaty, all signatories have an equal right to exploit the national resources of the archipelago of Svalbard (Spitsbergen), but the problem arises over whether this applies to the surrounding sea areas. Both these issues have major strategic implications. Apart from examining the complicated management of the Barents Sea, *Marine Management in Disputed Areas* offers a detailed analysis of two highly sensitive legal disputes. A certain amount has been written about these cases in the past, but this is the first detailed account in English. Although both authors are lawyers, and they discuss the problems from a non-partisan legal viewpoint (also suggesting possible solutions), the discussion is kept within its proper political and economic context, while simultaneously offering an informative account of the geography, oceanography and resource potential of these areas.

Dr Robin Churchill is Senior Lecturer in Law at Cardiff Law School, where he teaches international and EC law. He is particularly interested in maritime and environmental affairs. He has both researched and written extensively in this area. Geir Ulfstein is Senior Lecturer in Law at the Scandinavian Institute of Maritime Law, Oslo, where he teaches maritime law. His main research interests lie in international law and law of the sea, and he has researched and written extensively in these areas.

Ocean management and policy series
Edited by H.D. Smith

Development and social change in the pacific islands
Edited by A.D. Couper

Marine mineral resources
Fillmore C.F. Earney

Advances in the science and technology of ocean management
Edited by Hance D. Smith

The development of integrated sea-use management
Edited by Hance D. Smith and Adalberto Vallega

World ocean management
H.D. Smith and C.S. Lalwani

Marine management in disputed areas

The case of the Barents Sea

Robin Churchill and Geir Ulfstein

ROUTLEDGE

London and New York

First published 1992
by Routledge
11 New Fetter Lane, London EC4P 4EE

Simultaneously published in the USA and Canada
by Routledge
a division of Routledge, Chapman and Hall, Inc.
29 West 35th Street, New York, NY 10001

© 1992 Robin Churchill and Geir Ulfstein

Typeset in Times by
NWL Editorial Services, Langport, Somerset

Printed and bound in Great Britain by
Mackays of Chatham PLC, Chatham, Kent

All rights reserved. No part of this book may be reprinted or reproduced or utilised in any form or by any electronic, mechanical, or other means, now known or hereafter invented, including photocopying and recording, or in any information storage or retrieval system, without permission in writing from the publishers.

British Library Cataloguing in Publication Data
A catalogue record for this book is available from the British Library.

ISBN 0-415-03811-1

Library of Congress Cataloging in Publication Data
Churchill, R.R. (Robin Rolf)
Marine management in disputed areas: the case of the Barents Sea / Robin Churchill and Geir Ulfstein.
p. cm.
Includes bibliographical references and index.
ISBN 0-415-03811-1
1. Barents Sea – International status. 2. Marine resources conservation – Law and legislation – Barents Sea. 3. Marine resources conservation – Law and legislation – Norway – Svalbard. 4. Territorial waters. 5. Economic zones (Maritime law). 6. Norway – Boundaries – Soviet Union. 7. Soviet Union –Boundaries – Norway.
I. Ulfstein, Geir, 1951– . II. Title.
JX4084.B35C48 1992
341.4′48′0268470481–dc20 91–37180
CIP

Contents

Illustrations

MAPS

LIST OF FIGURES

Preface

The aims of this book are set out in detail at the beginning of Chapter 1. Very briefly and broadly, they are to provide a legal analysis of marine resource management in the Barents Sea. To some people the Barents Sea may seem a remote and obscure region, scarcely worthy of a full-length study. Such a view would be mistaken. As Chapter 1 explains, the Barents Sea is rich in resources, both actual and potential, and of great military and strategic significance. Moreover, because of the Treaty of Spitsbergen, which regulates the legal status of that archipelago and to which more than forty states are party, the Barents Sea is of very much more than purely local significance to its two riparian states, Norway and the USSR.

Whilst this book was in the press, the USSR ceased to exist. Russia has now succeeded the USSR as a riparian state of the Barents Sea. References to the USSR in this context should now be read as referring to Russia.

This book is a joint work in the full sense of that phrase. The book was jointly conceived and planned, and while naturally each chapter was individually drafted, each chapter was reworked in the light of the other author's comments and criticisms.

In writing this book we have received much help from various quarters and therefore we would like to thank the following: Dr Hance Smith, who invited us to write this book for the series that he is editing; our publishers, in particular Mr Alan Jarvis, for their patience and understanding as we missed various deadlines; Mrs Dawn Morgan, for coping so cheerfully with all the typing involved in producing this book; Mr Alun Rogers of the Cartographic Unit of the Department of Maritime Studies and International Transport, University of Wales College of Cardiff, for drawing the maps that appear in this book; and Mr Neil Phillips for research assistance.

Robin Churchill and Geir Ulfstein
Cardiff and Oslo
June 1991

1 Introduction

AIMS AND SCOPE OF THIS BOOK

The broad aim of this book, as one in a series partly concerned with the management of particular seas, is to look at the management of the natural resources of the Barents Sea. As the object of a study in regional marine resource management, the Barents Sea (see Map 1.1) is unusual in a number of respects. First, it has only two riparian states, Norway and the USSR, which for most of the time since 1945 have belonged to opposing military alliances. Second, the development of a proper regime for managing the natural resources of the Barents Sea has been greatly affected and hindered by two long-standing unresolved legal questions. The first of these is the location of the boundary between the maritime zones of Norway and the USSR: without an agreed boundary the proper management of resources is clearly made more difficult, and there is obviously more potential for conflict and friction between the riparian states. The other important unresolved legal question is whether the Treaty concerning the archipelago of Spitsbergen of 1920 (with its regime of equality of treatment for all its parties in the exploitation of the resources of Svalbard)[1] extends to the maritime zones of Svalbard beyond its territorial sea. If the Treaty does so extend, then all forty or so states party to it will have an equal right to exploit the resources of those zones: on the other hand, if the Treaty does not so extend, the exploitation of those resources will be the sole prerogative of Norway. Finally, the Barents Sea is unusual in the degree to which military and strategic issues influence the management of its natural resources: such issues also have a very important bearing on the possible resolution of the two legal questions just referred to.

Within the broad aims of this book just described, the authors have chosen to focus in particular on the two unresolved legal questions referred to above. They have done so for two reasons. First, these are the issues relating to the management of the natural resources of the

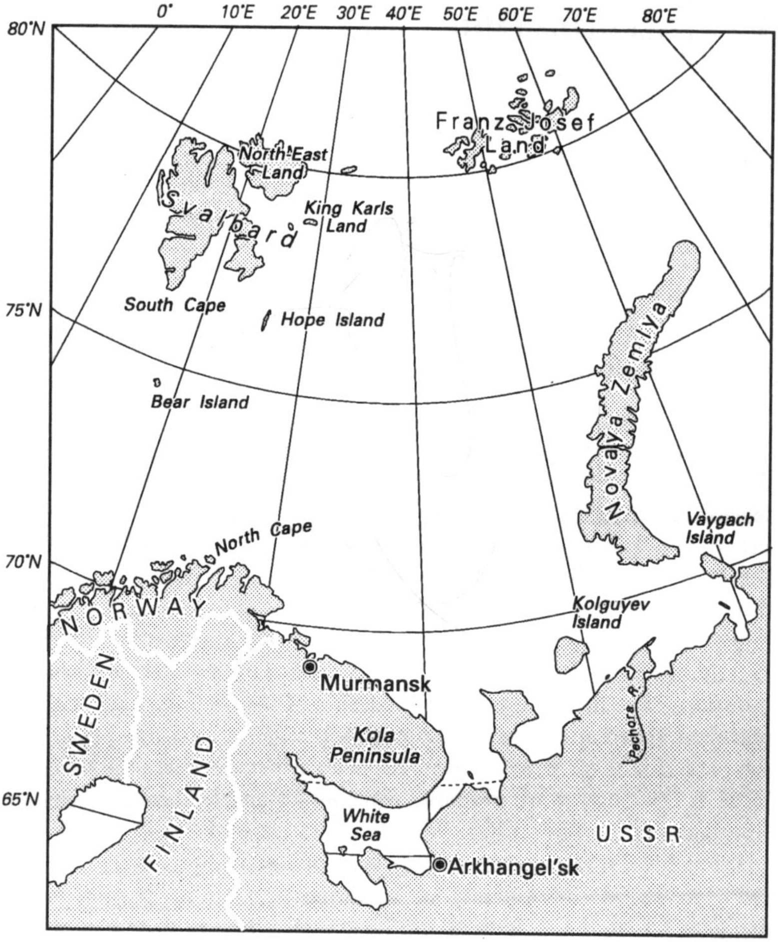

Map 1.1 The Barents Sea

Barents Sea which the authors, both of whom are lawyers, are most competent to discuss. Second, and perhaps more importantly, the authors feel that these two outstanding questions of the maritime boundary and the area of application of the Svalbard Treaty have not received the kind of extended, rigorous, detached legal analysis that they deserve: most of the existing literature on these questions (much of which is not in English) is either highly partisan and/or written not by lawyers but by political scientists. Chapters 2 and 3 attempt to remedy this deficiency by offering what is hoped will be considered a thorough, sustained and reasonably detached analysis of these two questions and some suggestions as to what the answers to these two questions might be.

The resources of the Barents Sea which have so far been the most important in practice are its living resources. The regime that has evolved for their management is analysed and subjected to critical evaluation in Chapter 4. While the focus, as in the preceding two chapters, is predominantly legal, the authors have attempted, as in those chapters, to place discussion of the legal issues in their proper political and economic context. The other major, though currently only potentially rather than practically important, natural resources of the Barents Sea are oil and gas. Chapter 5 discusses developments so far in the exploration and exploitation of these resources, and gives a brief account of the regime that currently governs these matters. This chapter also discusses the question of controlling pollution from future offshore development of oil and gas. This is the only discussion of pollution in the book. The reason why there is no further discussion is because the Barents is not a particularly polluted sea (although the same is not true of the adjoining Kola Peninsula, where pollution from its mining industries has become a serious problem, causing damage not only to the peninsula but also to adjoining land areas in Finland and Norway). Other reasons for limiting the discussion of pollution are that much of what pollution there is in the Barents Sea originates from outside the area and that there is no specific legal regime for pollution relating to the Barents (though of course many of the general marine pollution agreements do cover the Barents Sea).

The aim of the present introductory chapter is to give sufficient background information to facilitate understanding of the legal questions discussed in Chapters 2 and 3 and the management issues discussed in Chapters 4 and 5. This chapter begins with a description of the geography and oceanography of the Barents Sea. It then turns to give some idea of the enormous wealth of the living resources of the Barents Sea and an indication of the possibilities for exploiting oil and gas. This is followed by a brief account of the military and strategic significance of the Barents Sea, and of uses of the Barents Sea other than for military and resource purposes. The final part of the chapter outlines the framework of jurisdiction (that is, the different maritime zones) in the Barents Sea.

While much of the material in this book will be predominantly of interest in Norway and the USSR, the issues discussed do nevertheless have a much wider interest, not just to academic students of marine resource management, but also in a very real practical sense. All forty or so parties to the Svalbard Treaty, which include all the major developed states, have an interest, not to say a potentially very important economic stake, in the outcome of the question as to whether the Treaty applies to the maritime zones of Svalbard beyond its territorial sea, both in terms of fisheries interests and oil and gas potential. In addition, irrespective of the outcome of this question, a number of third states have vessels fishing regularly in the Barents Sea, and such states and their vessels are directly concerned with the regime for managing the living resources of the Barents Sea.

Before we turn to the subject-matter of this chapter outlined above, the necessary background information on the Barents Sea, its resources and uses, we must first seek to try to define the area and boundaries of the sea we are dealing with.

DEFINING THE BARENTS SEA

It is difficult to define seas with great precision, except for those seas that are virtually enclosed, like the Mediterranean and the Red Sea. Nevertheless, there is a fairly generally accepted definition of the Barents Sea, whose use is endorsed by the International Hydrographic Bureau. This definition is as follows. To the south the Barents Sea is bounded by the mainland coasts of Norway and the USSR, to the east by the large Soviet archipelago of Novaya Zemlya, and to the north by the archipelagos of Franz Josef Land and Svalbard, belonging to the USSR and Norway respectively. In the west the Barents meets the Greenland and Norwegian Seas, the conventional dividing line being taken as the line running from the South Cape (on Svalbard), through Bear Island (Norwegian) to the North Cape on the Norwegian mainland. Thus defined, the Barents Sea is about 1.4 million square kilometres (about 542,000 square miles) in area.[2]

OCEANOGRAPHY

The Barents Sea is comparatively shallow, only about half being deeper than 200 metres. The average depth is 229 metres.[3] South of Bear Island the sea-bed plunges to about 500 metres in Bear Island Trench, and there are a number of smaller troughs or trenches – the South Cape, North and North-east Trenches. Geologically speaking, the whole of the bed of the Barents Sea is continental shelf, part of the world's largest continuous continental shelf, extending from the Barents Sea in the west to Siberia and the Chukchi and Beaufort Seas in the east, and northwards several hundred miles north from Norway, the USSR, Alaska and Canada. This continental shelf also extends somewhat to the west of the western boundary of the Barents Sea described above.[4]

Thanks to the Gulf Stream, the Barents Sea and its coasts are significantly less cold than all other areas at the same latitudes. As a result, there is much less ice than in other sub-Arctic seas. The south-west part of the Barents Sea, from just south of Bear Island to the eastern end of the Kola Peninsula, is free of ice all the year round. The remainder is covered, for at least part of the year, with ice in various forms (drift-ice, pack-ice, ice ridges and icebergs), although in the summer the whole of the Barents south of a line between Svalbard and Novaya Zemlya is free of ice and in September, the best of the summer months, it is possible to sail to Franz Josef Land. Most of the ice in the Barents Sea is of local origin, though there is some movement of ice from the Polar Basin and Kara Sea into the Barents Sea. Most ice is less than one year old, and apart from ice ridges (which on average have ice keels of 12 m) and icebergs, does not exceed one metre in thickness.[5]

Tidal amplitude varies greatly, from 13 feet (4 m) at North Cape and 30 feet (9m) at Strait Gorlo on the White Sea, to 5 feet (1.5 m) at Svalbard and 2.6 feet (0.8 m) in the vicinity of Novaya Zemlya.[6]

The climate is sub-Arctic, with winter air temperatures averaging –25°C in the north and –5°C in the south-west, and summer air temperatures averaging 0°C and 10°C, respectively. Annual precipitation is 20 inches (500 mm) in the south, but only half that in the north.[7] The Barents lies in the path of warm cyclones from the North Atlantic and cold anti-cyclones from the Arctic, producing very unstable climatic conditions and some of the most storm-ridden stretches of sea in the world.[8]

GEOGRAPHY

The Barents Sea takes its name from the Dutch explorer, Wilhelm Barents, who made a number of expeditions to the area in 1594–6. The Barents Sea was, however, known and used long before the voyages of its eponymous explorer. It is likely that prehistoric man lived on its shores, as rock carvings have been discovered at Alta, some miles south-west of the North Cape, which are believed to be 5–6,000 years old. In historic times, there is little doubt that the Barents was sailed on extensively by the Vikings, who are thought to have known about Svalbard in the twelfth century. From the thirteenth century onwards there was a vigorous seaborne trade between North Norway and Northern Russia. In the sixteenth century came the Dutch, who 'rediscovered' Bear Island and Svalbard in 1596. In the seventeenth and eighteenth centuries there was large-scale hunting of walrus and whales in the seas around Svalbard by British, Dutch, French, Hanseatic, Danish, Norwegian, Russian and Spanish vessels. From the nineteenth century onwards, as we will see below, the Barents Sea and its bordering coasts have been subject to a wide variety of economic and other activities.

We will now turn to say something about each of these bordering coasts, mainland and insular, beginning with the mainland coast of Norway and moving clockwise around the Barents Sea.

The coast of Norway bordering the Barents Sea runs from the terminus of the Soviet–Norwegian land border on the shores of Varangerfjord westwards to the North Cape. This coastline, often comprising steep cliffs, is deeply indented, being penetrated by a number of large fjords. Unlike most of the Norwegian coastline, the easterly part of the Barents coast, the Varanger and Nordkinn peninsulas, is almost wholly devoid of offshore islands: only in the vicinity of the North Cape (which itself is on an island) do islands start to become more numerous. This part of Norway is thinly populated. The entire population of the county of Finnmark, which includes not only the Barents Sea coast but also the coast some 130 miles south-west of the North Cape, as well as the adjacent inland areas up to the frontiers with the USSR and Finland, is only 74,000. Most of this population lives in small scattered, isolated communities on the coast. Fishing forms the backbone of the economic activities of these communities, although around Kirkenes, close to the Soviet border, there is also iron-ore mining. Inland from the coast live the Lapps, whose livelihood is largely based on reindeer: there are also some Lapps living on the coast who make a living from sea fishing.

Some 220 nautical miles north of the North Cape lies Bear Island. This small, isolated, pear-shaped island belongs to Norway, but legally is part of Svalbard rather than mainland Norway, the significance of which will become apparent shortly. Bear Island is only 178 square kilometres in area. It has no permanent population, although there is a weather station which is staffed all the year round by teams of meteorologists. Otherwise, other scientists spend time on the island, as does the occasional hunter.

Some 130 nautical miles north of Bear Island lies the South Cape, the most southerly point of the Svalbard archipelago. The latter comprises a number of large islands – Spitsbergen (formerly called West Spitsbergen), North-east Land, Barents Island and Edge Island – together with a large number of smaller islands, islets and rocks, making a total area of 62,400 square kilometres (roughly the size of Belgium and the Netherlands combined). The archipelago belongs to Norway, but as we shall see in more detail in the next chapter, the Treaty of 1920 that gave Norway sovereignty over Svalbard also provided that the nationals of the other forty or so states party to the Treaty were to enjoy the same rights as Norwegian citizens in relation to economic activities, including mining, hunting and fishing. About two-thirds of Svalbard is permanently covered by ice and in several places glaciers reach the sea. Land forms are largely Alpine, and much of the coast is indented by fjords. There are no indigenous inhabitants on Svalbard. Most of the thousand or so Norwegians who live on Svalbard are engaged in coal-mining, producing about 400–450,000 tonnes a year. Coal-mining, which began in 1899, is the major, and indeed – apart from a few individuals hunting animals for their furs, and tourism in the summer months – the only, economic activity. The Russians also engage in coal-mining, producing about the same amount of coal as the Norwegians, but with two or three times the population. The USSR is the only party to the Treaty currently exercising its rights to engage in economic activities on the same footing as Norway,[9] although since the late 1960s oil companies from other states parties (and Norway) have been exploring for oil, although so far without making any commercial finds. Other minerals found on Svalbard, though again not (yet) in commercial quantities, include iron, lead, copper, zinc, marble, gypsum and asbestos. There is also a certain amount of scientific research on Svalbard, undertaken by scientists from many of the states party to the 1920 Treaty. Finally, the waters off Svalbard are rich in fish stocks and support a large commercial fishery: none of the vessels engaging in this fishery operate from Svalbard, however. More is said about this fishery on pp. 9–12, when discussing the resources of the Barents.

To the east of Svalbard lies another archipelago, Franz Josef Land. This comprises some 191 islands, arranged in three main groups, occupying an area of 16,134 square kilometres. Franz Josef Land consists mainly of low-lying plateaux, 85 per cent of whose surface is permanently covered by ice. The archipelago became part of the USSR in 1926. Apart from meteorologists working at a number of permanent weather stations, there are no people living on Franz Josef Land.

Some 200 nautical miles south of Franz Josef Land lies the northern end of the large Soviet archipelago of Novaya Zemlya. The archipelago consists of two very large islands, Severny and Yuzhny, separated by the very narrow Matochkin Shar strait, together with several smaller islands. The archipelago extends for about 2,000 kilometres along a north-east–south-west axis and is about 82,600 square kilometres in area. Novaya Zemlya is a continuation of the Ural Mountains system, and therefore is largely mountainous. Over a quarter of its land area is permanently covered by ice. The only people living on Novaya Zemlya are a number of scientists working at meteorological stations. The original indigenous population, some 400 Samoyeds (or Nenets), were forcibly removed during the 1960s when Novaya Zemlya was used for the atmospheric testing of nuclear weapons. Testing has recently been resumed, although it is now underground.[10]

Separated from the southern end of Novaya Zemlya by the fifteen-mile-wide Kara Gates Straits is the large island of Vaygach, which in turn is separated from the mainland by the much narrower Iugorskii Shar Strait. From this point the generally low-lying and sparsely inhabited Soviet mainland coast, ice-locked for much of the year, runs westwards to the Kanin Peninsula, the eastern entrance to the White Sea. North-east of the Kanin Peninsula lies the large island of Kolguyev (about 3,500 square kilometres) the only island of any size, apart from the archipelagos forming its periphery, in the Barents Sea. There is a fair-sized oilfield on the island from which commercial production has recently begun (see below, p. 13).

The White Sea is excluded from the definition of the Barents Sea given earlier. Thus the final stretch of coastline to be considered in this brief geographical overview of the Barents Sea is the Kola Peninsula, lying between Cape Sviatoi Nos, the western entrance point to the White Sea, and the Soviet–Norwegian frontier. The Kola Peninsula is by far the most heavily populated and most economically active and diverse of the Barents Sea's coasts. Much of the Kola Peninsula coastline is fairly straight, but towards its western end it is penetrated by a number of fjord-like indentations which provide several excellent sheltered harbours which are free of ice all the year round. These harbours, including that of Murmansk, are the home of the USSR's Northern Fleet, its largest navy, about which more is said below in the section on the strategic significance of the Barents Sea, and the USSR's largest fishing fleet. Apart from these maritime activities, the Kola Peninsula is also the site of mining for nickel, apatite and iron ore. Altogether these economic activities support a population of about one million on the Kola Peninsula, of which about half live in or very close to Murmansk.

NATURAL RESOURCES

The Barents Sea has one major actual natural resource and one potential natural resource of significance. The former is fish: the Barents Sea has long supported one of the world's major commercial fisheries. The latter is submarine deposits of oil and gas, about whose extent there is still considerable uncertainty. Each of these resources will be considered in turn.

Fish

The meeting of warmer waters from the Atlantic carried by the Gulf Stream with colder waters from the Arctic in the western part of the Barents Sea produces a phenomenon known as a polar front, whereby rich nutrients, such as phosphorus and nitrogen, are mixed with more habitable waters to support a vast quantity of phytoplankton on which lower forms of marine life feed.[11] The latter in turn form the food for large numbers of fish, which in turn are the food supply for seals, walrus, some kinds of whales and polar bears.

The earliest large-scale exploitation of living resources in the Barents Sea was not for fish but for marine mammals. As mentioned above, from the early seventeenth century whales, walrus and seals were hunted by considerable numbers of vessels from various countries. By the end of the eighteenth century the resources were largely exhausted through over-exploitation, and these activities came to an end. In the later part of the nineteenth century commercial whaling began again, prosecuted mainly by the Norwegians and Russians, but by the early twentieth century the resource was again badly depleted through over-exploitation. Since then there has been some relatively limited catching of minke whales by Norwegian whalers, but this came to an end in 1986, with the adoption of a ban on commercial whaling by the International Whaling Commission (see Chapter 4 below). The commercial catching of seals also revived in the nineteenth century, prosecuted largely by the Norwegians and Russians. Catches in the Barents and White Seas reached a zenith of just over 200,000 animals a year around 1950. Since then catches have declined, particularly since the mid-1980s the demand for seal skins has substantially decreased, mainly as the result of the activities of certain environmental groups, and aided by bans on the import of seal skins by the USA and the EEC. In 1989 about 43,000 seals were killed.[12]

Large-scale commercial fishing began in the Barents Sea in the late nineteenth century with the development of larger and stronger fishing vessels, better gear, and better ways of preserving and marketing the catch. In recent years the most commercially important species fished in the Barents Sea have been cod, haddock, saithe, redfish, capelin, herring and shrimp. Something will now be said about the stocks of each of these fish in the Barents.[13] It is also helpful to bear in mind that by no means all of these stocks are confined to the Barents Sea during their life cycle.

As far as cod is concerned, there are in fact two distinct stocks in the Barents, the North-east Arctic cod and the Polar cod. The former, historically the most commercially valuable stock in the Barents, inhabits the Barents Sea between North Norway and Svalbard and along the mainland coasts of Norway and the USSR (and also part of the North Norwegian coast outside the Barents), its northern and eastern limits effectively corresponding to the edge of the summer ice. Spawning takes place in March–April close inshore, mainly around the Lofoten islands (i.e. outside the Barents), whence developing pelagic eggs drift northwards to arrive as fingerlings over a wide area of the western Barents by late summer. The juveniles stay in the Barents, although migrating seasonally according to the hydrographic regime, until they are mature, when they return to the Lofoten islands to spawn. After spawning they return to the Barents. An inshore net and line fishery for cod has existed in northern Norway throughout recorded history. Offshore trawling started to develop at the end of the nineteenth century, and has increased steadily throughout the twentieth century. By 1955 the stock was fully exploited, but fishing continued to increase, reaching a peak catch of 1,197,000 tonnes in 1969. Not surprisingly, as a result of this over-fishing, and also certain environmental factors (explained further in Chapter 4), catches thereafter declined, reaching a low of 278,000 tonnes in 1984. Since then catches have increased somewhat, and in 1987 (the best year since 1984) the catch was 523,000 tonnes.[14] In spite of this modest increase, the size of the North-east Arctic cod stock still remains at a very low level (as explained further in Chapter 4).

The Polar cod, preferring colder water than the North-east Arctic cod, spends its life mainly in the eastern part of the Barents. It spawns in December–February in the South-eastern Barents and then gradually migrates northwards until September, when it returns to the spawning area. Fishing for Polar cod only began on any scale in the late 1960s. Subsequently catches have fluctuated between 130,000 and 350,000 tonnes a year: in 1987 (the most recent year for which figures are available) the catch was 145,000 tonnes.

The distribution and life cycle of the North-east Arctic haddock are similar to those of the North-east Arctic cod, as is the history of its exploitation, except that it has been less heavily over-fished. Catches reached a peak of 187,000 tonnes in 1972 and thereafter declined to about 30,000 tonnes a year in the early 1980s. Since then catches have improved significantly, and in 1987 the catch was 113,000 tonnes.

The North-east Arctic saithe migrates from the Barents and Northern Norway to spawn off western Norway. The eggs and fry are then carried northwards, back to Northern Norway and eventually the Barents. Catches of this stock (some of which are made outside the

Barents) peaked at around 220,000 tonnes in the early and mid 1970s. Since then they have declined as a result of over-fishing, and in 1985–8 averaged about 96,000 tonnes a year.

The Barents Sea stock of the North-east Arctic redfish spawns off the northern coast of Norway. As they grow, redfish migrate over the whole of the western and southern parts of the Barents. In the period 1962–77 catches averaged about 19,500 tonnes a year. More recently catches have declined and in the period 1985–8 the average annual catch was about 5,300 tonnes.

Prior to the late 1960s the Barents Sea was an important nursery area and feeding ground for the Norwegian spring spawning stock of the Atlanto-Scandian herring, which spawned off the west and north coasts of Norway, and supported a sizeable herring fishery. As the result of heavy over-fishing in the 1960s, however, the stock declined to an extremely low level, and catches of herring in the Barents Sea during the 1970s and early 1980s were insignificant. More recently, there has been some revival, and in the period 1985–8 catches averaged around 172,000 tonnes a year.

Capelin spawn along the North Norwegian and Murman coasts in March and April. They then migrate northwards, spending the summer in the northern part of the Barents, before moving south to the central part of the Sea in autumn and returning to the coast in winter. Before the mid 1960s fishing for capelin was on a very modest scale, but from the mid 1960s catches increased rapidly, reaching 1.5 million tonnes by the early 1970s and nearly 3 million tonnes in 1977. From then until 1983 catches averaged just under 2 million tonnes a year. Thereafter catches declined markedly as a result of over-fishing and environmental factors, and because of the poor state of the stock fishing for capelin was prohibited in 1987 and was not permitted to begin again until 1991. Although catches of capelin were very high in the 1970s and early 1980s, the fishery was not as commercially important or valuable as the cod and other whitefish fisheries described above, because capelin, a small and oily fish, is used mainly for reduction to meal and oil, which produces a much lower income than when fish are sold for human consumption.

Finally, the Barents supports an important shrimp fishery. Several thousand tonnes (37,000 in 1987) of this high-value crustacean are caught every year.

Overall in the last few years the total catch in the Barents Sea has been of the order of 1–1.5 million tonnes a year, compared with peak catches in the mid and late 1970s of about 4–4.5 million tonnes a year (of which two-thirds was capelin). At that time the catch was about six or seven per cent of the total world marine fish catch (even though the

Barents occupies only about 0.3 per cent of the area of the world's seas and oceans): now the catch represents only about 1.5–2 per cent of the world total. Norway and the USSR have always been the principal nations fishing in the Barents (and their Barents Sea catches represent a fair proportion of their total fish catches – currently around ten and five per cent, respectively), but the fishermen of a number of other European countries also fish in the Barents, indeed some such countries have a long history of participation in the Barents Sea fisheries. Since the establishment of 200-mile fishery and economic zones in 1977 (discussed in the final section of this chapter), the level of third country fishing in the Barents has declined drastically: Norway and the USSR now take over ninety per cent of the catch. Other countries still fishing in the Barents (in rough order of importance) are Spain, the Faroes, Germany (both the former East and West), France, the United Kingdom and Denmark.

Oil and gas

The bed of the Barents Sea has long been suspected of containing oil and gas in considerable quantities. These expectations have been based on existing (if currently incomplete) geological knowledge of the bed of the Barents Sea,[15] on the fact that the south-eastern corner of the Barents is adjacent to the prolific onshore Timano–Pechora Basin oilfield, and on the fact that the Barents Sea is part of the same continental shelf as that of Prudhoe Bay and the Beaufort Sea, where large quantities of oil have been found. There is thus considerable oil industry interest in the Barents, not least because it is one of the few areas left in the world offering hope of substantial oil and gas finds which so far is relatively unexplored.

Norway began initial exploration off its northern coast, off Troms, just outside the Barents, with seismic surveys from 1975 onwards. Exploratory drilling began in 1980. Subsequently, both seismic surveying and exploratory drilling have been extended northwards and eastwards into the Barents Sea proper. Oil has been found, but so far not in sufficient quantity or at the necessary pressure to allow commercial production, although the find made by Shell in Lopparyggen in 1989 may be more promising. Greater quantities of gas have been found, but because of the distance of the area from the markets for gas and the price of gas, commercial production is not (yet) feasible.

The USSR began seismic surveys in the Barents Sea in 1978 and exploratory drilling in 1982. The latter has been concentrated in four areas – in the Severo-Kildinskaya area, close to the disputed continental shelf area; on the Murmansk High, off the Kola Peninsula; on Goose Bank, west of the southern island of Novaya Zemlya; and off the mouth of the Pechora River, in the south-eastern Barents.[16] Oil has been found in this last area, and gas on the Murmansk High, though in neither case is it thought yet in sufficient quantities to justify commercial production. On the other hand, in August 1989 the USSR discovered a huge gas field in the Shtomanovskaya structure, which is not in any of the areas just mentioned, but about 200 miles west of Novaya Zemlya and 350 miles north-east of Murmansk. Initial estimates suggest that the field could contain three to four trillion cubic metres of gas, which if proved correct, would make it the fourth largest gas field in the USSR and by far its biggest offshore gas field.[17] It is unlikely that the field will be developed until the twenty-first century. It should also be mentioned that oil and gas were found on Kolguyev Island in 1983 and commercial production of oil began in 1987, which it was hoped would reach about 0.5 million tonnes a year in 1990. Although this oilfield is on land, it is thought that it might extend out under the sea.

While the above developments do not yet prove that the Barents Sea will turn out to be a major offshore oil and gas province, they do suggest that hopes that the Barents may hold considerable quantities of oil and gas are not misplaced.

MILITARY AND STRATEGIC SIGNIFICANCE

The military and strategic significance of Northern waters, and particularly the Barents Sea, has been the subject of much discussion by military analysts, political scientists and others.[18] In view of the apparent ending of the Cold War and political changes in the USSR, much of this discussion, particularly its more speculative aspects, must now be regarded as being out of date. Because of the rapidity with which changes in East–West relations and the political situation in the USSR are occurring, it seems idle to engage in any extensive speculation about the precise military and strategic significance of the Barents Sea in the future, even if the authors were competent to undertake such a task, which they are not. Instead, it may be more useful briefly to set out certain well-established facts and factors concerning the military situation in the Barents Sea which are likely to remain constant and valid for the next few years.

Since the early 1960s the USSR has built up an enormous naval base, together with an associated complex of air bases and other military installations, on the Kola Peninsula. This base is now the largest naval base in the world. The fleet housed there, the Northern Fleet, is the largest of the USSR's four naval fleets and contains a majority of the USSR's submarines (including those carrying ballistic missiles) and a large proportion of its most powerful surface warships. The Northern Fleet is the USSR's most important fleet, not only because of its size but also because of the location of its base: the USSR's other three fleets, based in the Baltic, the Black Sea and the Sea of Okhotsk, suffer the disadvantage that their home ports are blocked by ice for part of the year and/or that to reach the great oceans of the world they must pass through narrow straits bordered by Western powers. By contrast, the Northern Fleet enjoys both a base which is ice-free all the year round and fairly unconstricted access to the world's oceans through the 350-mile-wide Svalbard Passage between North Norway and Svalbard, with the added advantage that the land on the northern side of this Passage, Svalbard, even though belonging to a member of NATO, is, as a result of the 1920 Svalbard Treaty, demilitarised. For these reasons the Northern Fleet is likely to remain the most important of the USSR's four fleets, and therefore, assuming that naval power continues to form an important component of the USSR's armed forces, the Kola Peninsula is likely to continue to be a militarily most important area. This last assumption, about the continuing importance of naval power to the USSR, seems valid. The trend, especially in East–West arms negotiations, is towards a reduction in arms and manpower on land, rather than at sea. More generally, in spite of the ending of the Cold War, the USSR is likely to continue to be, as it has been throughout its history, preoccupied by its relative defensive vulnerability and the need therefore to maintain a strong defensive capability. In particular, as far as Northern waters are concerned, it is not forgotten that during the Civil War after the 1917 Revolution, Western states provided aid to the Whites through the ports of Murmansk and Arkhangel'sk.

The Barents Sea is not only a transit area for the USSR's Northern Fleet. With developments in the range of submarine-launched ballistic missiles, it has also since the mid 1970s been an operating area for Soviet submarines. Nuclear-armed missiles launched from submarines in the Barents Sea can reach most of North America, Western Europe and China. As long as long-range nuclear weapons remain an important part of the USSR's arsenal, and there is no reason to suppose that they will not, the Barents Sea will continue to be an important operating area for submarines carrying ballistic missiles, even if only for defensive purposes. The particular advantages of the Barents Sea for these purposes are: first, that it is not necessary for such submarines to move far from their home base into potentially hostile waters beyond; and, second, that parts of the Barents Sea offer ice cover for submarines, thus making their detection, whether from the air or by underwater listening device, more difficult.

Because of their importance and their relatively exposed and vulnerable position, concentrated in a small area only 100 kilometres from a NATO country (Norway), the bases on the Kola Peninsula have made the USSR extremely sensitive and concerned about any non-Soviet activities either in land areas adjacent to the Kola Peninsula or in the adjoining Barents Sea. As long as the Kola Peninsula retains its pre-eminent position as a naval base, these Soviet sensitivities and concerns can be expected to continue.

So far we have looked at the military and strategic significance of the Barents Sea solely from the point of view of the USSR. But the fact that the Barents is of such great significance to the USSR has meant, of course, that it has also been of great military and strategic interest to other states, especially the USSR's neighbour and fellow Barents Sea riparian, Norway, and its allies in NATO. In the past the USSR's Northern Fleet was regarded as a threat to the North Atlantic sea-lanes between the USA and Western Europe, along which the USA would get reinforcements to Europe in the event of war, as well as a more direct threat to the reinforcement of Norway in such an event. To counter this perceived threat, the USA developed a strategy whereby it would have the capability to take on and check the Soviet navy in the north Norwegian and Barents Seas, before that navy could cut the North Atlantic sea-lanes of communication. Whether, with the ending of the Cold War, the USA still intends to maintain this strategy, the authors have not been able to discover. What seems certain, however, is that the USA and its NATO allies will continue to keep developments on the Kola Peninsula, and shipping movements in and out of its ports, under constant surveillance.

Turning from the strategy of NATO in general to Norway in particular, as a neighbour of the USSR Norway has been both more aware of and more sensitive to the military significance of the Barents Sea for the USSR than other NATO members. The policy of successive Norwegian Governments since the Second World War has been to try to keep tensions in the area low (for example by not permitting foreign troops to be stationed in Norway and by insisting that NATO exercises with visiting foreign troops in northern Norway are held well away from the frontier with the USSR) and seeking to respect the legitimate security interests of the USSR, while at the same time deterring the USSR from any aggressive actions towards itself through its membership of NATO. There seems no reason to suppose that this policy will change in the immediate future.

To sum up: notwithstanding the ending of the Cold War, the Barents

Sea is and is likely to remain an area of great military and strategic significance, primarily because the Kola Peninsula is the USSR's largest and most important naval base and because the Barents Sea is both an important transit area for its Northern Fleet and a major operating area for its submarines carrying nuclear-armed ballistic missiles. As will be seen in the following two chapters, the military and strategic significance of the Barents Sea has an important bearing on the two main legal issues discussed in this book.

OTHER USES OF THE BARENTS SEA

We have now considered the major uses of the Barents Sea – for fishing, oil and gas exploitation and military purposes,[19] but there are a number of less important uses which must be briefly mentioned. These include shipping, overflight and research.

As regards the first of these, apart from fishing vessels, vessels involved in offshore oil and gas activities (for example, seismic surveying) and military vessels, all of which have been mentioned or alluded to in the preceding sections, a number of other shipping activities take place in the Barents Sea. These include: the transport of coal from Svalbard to Norway and the USSR; inshore passenger and goods traffic along the Norwegian coast (largely local in character); the transport of goods between Murmansk and Arkhangel'sk and non-Soviet ports around northern Norway; traffic along the USSR's Northern Sea Route, the western end of which is in the Barents and which outside the summer months is kept open by a fleet of ice-breakers; and cruise ships carrying tourists to the North Cape, Svalbard and (from the USSR) Franz Josef Land.

As well as transport on the surface of the Barents Sea, passengers and goods are also transported above it by aircraft. The main air routes in the Barents are between Svalbard and North Norway, and Svalbard and the USSR. It may also be that longer routes, for example between Western Europe and Japan and between Moscow and North America, also cross part of the Barents.

Finally, a considerable amount of research takes place in the Barents Sea. Apart from research connected directly with fisheries and oil activity, which is carried out mainly by Norway and the USSR, there is also a good deal of pure oceanographic research, conducted not only by Norway and the USSR, but also by a number of other states including in particular Germany and the UK. There is a reasonable amount of co-ordination of fisheries research through the International Council for the Exploration of the Sea, and it may be that in time there will be some co-ordination of other research through the recently established International Arctic Science Committee. In addition, Norway and the USSR co-ordinate some of their research under their Agreement concerning Technical and Scientific Co-operation in the Arctic and Northern Areas, which was signed in 1988.

THE FRAMEWORK OF JURISDICTION

Legally, the sea is divided into various zones. These zones determine what kind of activities different states can carry out in particular areas and the competence of different states to regulate marine activities in those areas. The international law governing this zonal division of the sea is contained in a mixture of rules of customary international law and treaties – the four Geneva Conventions of 1958 and the United Nations Convention on the Law of the Sea 1982 (which is not yet in force). The aim of this section is to give a brief account of the various jurisdictional zones at sea, and explain how they apply in the Barents Sea.

The various zones are measured, not from the coast as such, but from what are known as baselines. The baseline is normally the low-water line along a particular stretch of coast. However, as an exception, along coasts which are deeply indented and/or fringed with islands, instead of using the low-water line as the baseline, a state may draw straight baselines connecting the outermost points on the coast and/or islands, provided that such lines follow the general direction of the coast. In practice straight baselines have been drawn along most of the coasts of the Barents Sea. Norway has drawn straight baselines along the whole of its mainland coast in the Barents,[20] around Bear Island,[21] and along the south and west coasts of the Svalbard archipelago.[22] The USSR has drawn straight baselines along much of its mainland coast in the Barents, around Novaya Zemlya and Vaygach Island, and along the coasts of some of the islands in the Franz Josef Land archipelago.[23] Another important exception to using the low-water line as the baseline is to draw a closing line across the mouth of a bay, either where the bay is less than twenty-four miles in width or where the bay is recognised as an historic bay. In the Barents Sea the USSR has drawn closing lines across two bays on the grounds that they are historic bays. These are the White Sea and the Bay of Cheshsk (immediately east of the Kanin Peninsula).[24]

Waters on the landward side of baselines are known as internal waters. Such waters are under the sovereignty of the coastal state. Other states have no rights to enter or utilise internal waters except in two situations: first, where a right may have been granted by treaty (for example, to enter and use the ports of the coastal state); second, under article 5(2) of the 1958 Convention on the Territorial Sea (and under article 8(2) of the 1982 UN Convention) where straight baselines are drawn enclosing as internal waters areas which had not previously been considered as such, there is a right of innocent passage (a concept explained below) for other states through such internal waters. This latter provision applies in respect of the straight baselines drawn in the Barents Sea in 1985 by the USSR, which (unlike Norway) is a party to the 1958 Territorial Sea Convention.

The first zone seawards from the baseline is the territorial sea. This zone, like internal waters, is subject to the sovereignty of the coastal state. Unlike internal waters, however, other states do enjoy one general right, both under customary law and the conventions, in the territorial sea. This is the right of innocent passage. What this means is that foreign ships can sail through the territorial sea of the coastal state provided that their passage is not prejudicial to the peace, good order or security of the coastal state: the 1982 Convention lists certain activities (e.g. weapons practice, serious pollution, fishing, research or survey activities) which are deemed to be prejudicial in this way. A ship which is exercising the right of innocent passage is subject to the jurisdiction of the coastal state: thus it must comply with the coastal state's regulations relating to such matters as navigation routes and pollution prevention. In straits used for international navigation lying in the territorial sea both the Convention on the Territorial Sea and the UN Convention strengthen the navigational rights of foreign states. These provisions will not be further considered here, however, because while the Barents Sea does, at its eastern end, contain a number of straits lying in the territorial sea, none of these straits is used for international navigation.

The maximum permissible breadth of the territorial sea was for many years controversial. The UN Convention of 1982 establishes a maximum breadth of twelve miles, and while the Convention is not yet in force, it is now almost certain that the same limit exists in customary international law. The USSR has long claimed a twelve-mile territorial sea, a claim having first been made by the Tzar in 1909.[25] Norway has claimed a four-mile territorial sea since 1812,[26] and although about three-quarters of all coastal states now claim twelve miles, it has not apparently contemplated extending its own claim to twelve miles.

The next zone beyond the territorial sea which is relevant in the Barents Sea is the continental shelf. The continental shelf, as a legal doctrine, developed in customary law following claims made by the USA and other states from 1945 onwards, a development that crystallised in the 1958 Convention on the Continental Shelf. Under this Convention, and the 1982 Convention, the continental shelf is an area of sea-bed and subsoil beyond and adjacent to the territorial sea where the coastal state enjoys exclusive sovereign rights to explore for and exploit the natural resources of the sea-bed and subsoil. These rights do not, however, affect the legal status of the waters above, which will be exclusive economic zone or high seas (terms explained below). In particular, in exercising its continental shelf rights (which include the right to erect installations to explore for and exploit the shelf's natural resources), a coastal state must not interfere unjustifiably with the rights which other states enjoy in the superjacent waters.

The 1958 Convention defines the outer limit of the continental shelf as being either the 200-metre isobath or, if it is possible to exploit the natural resources of the sea-bed beyond that point, 'to where the depth of the superjacent waters admits of the exploitation' of those resources. This latter definition, it will be observed, is a very flexible one, the outer limit of the continental shelf changing as the technology of offshore oil and gas exploitation develops, and therefore has been widely regarded as being unsatisfactory. Partly in response to these feelings of dissatisfaction, the 1982 Convention adopts a different definition of the outer limit of the continental shelf. Its article 76 defines the outer limit as being either 200 miles or the outer edge of the continental margin, whichever is the further. Where it is the latter which is the further, the Convention provides that for legal purposes the outer edge of the continental margin is *either* a line connecting points not more than sixty miles apart, at each of which points the thickness of sedimentary rocks is at least one per cent of the shortest distance from such point to the foot of the continental slope, *or* a line connecting points not more than sixty miles apart, which points are not more than sixty miles from the foot of the slope. In each case the points referred to are subject to a maximum seaward extent: they must be either within 350 miles of the baseline or within 100 miles of the 2,500 isobath. There is now a growing body of state practice utilising the 1982 Convention's definition, with the result that even though the Convention is not in force, its definition of the outer limit of the continental shelf is in the process of passing into customary international law. It should be noticed that under both the 1958 Convention and the 1982 Convention the continental shelf, and particularly its outer limit, is for legal purposes defined in a way quite different from the way the term is used by geologists and geomorphologists.

Although under the 1958 Convention (and 1982 Convention) there is no need for a state expressly to claim a continental shelf, since in law it has an automatic right to a continental shelf, both Norway and the USSR have (like practically all coastal states) formally claimed a continental shelf, in 1963[27] and 1968[28] respectively. The Norwegian claim of 1963 defined the outer limit of the Norwegian continental shelf as being 'as far as the depth of the superjacent waters admits of exploitation of natural resources', i.e. the second alternative definition of the 1958 Convention (to which Norway is a party). In 1985 this definition was amended by the Petroleum Law[29] to 200 miles or the edge of the continental margin, i.e. the definition of the 1982 Convention, which Norway has signed but not ratified. The Soviet legislation defines the outer limit of the USSR's continental shelf in terms identical to those of the 1958 Convention (to which the USSR is a party). Under the 1958 Convention's definition, and given the current state of offshore technology, the whole of the Barents Sea is, legally speaking, probably continental shelf. Under the 1982 Convention there is no doubt that this is the case.

Overlapping the continental shelf, at least as far as the sea-bed is concerned, is the exclusive economic zone (EEZ). The EEZ is a relatively recent concept in the law of the sea, being a product of the Third United Nations Conference on the Law of the Sea (whose outcome was the 1982 Convention). The Convention provides for the establishment of an EEZ 200 miles in breadth measured from the baseline. Since 1976 and while the Conference was still in session, a large number of states have claimed 200 mile EEZs. Because of the number of claims, and the absence of any significant protest to them, it is now certain that the EEZ has become part of customary international law.[30] The Convention provides that within its EEZ a coastal state enjoys various rights, all of which are essentially connected with the exploitation of the zone's natural resources. First, the coastal state has the right to exploit the non-living resources of the EEZ, such as oil and gas, sand and gravel. (It is in this respect that the EEZ overlaps with the continental shelf.) Second, the coastal state has the right to exploit the living resources of the EEZ. This right is, however, subject to various management duties, which are discussed in detail in Chapter 4. Third, the coastal state has the right to exploit the EEZ for other economic purposes, such as the production of energy from the waves, wind or currents. Fourth, the coastal state has the right to authorise and regulate the construction and operation of artificial islands, installations and structures, such as drilling rigs for oil and gas exploitation or wave barrages. Fifth, the coastal state has the right to regulate scientific research in its EEZ. It must normally give its consent to pure research by other states in its EEZ, but it may withhold its consent to resource-oriented research. Finally, the coastal state has various rights to control pollution in its EEZ. In particular, it can regulate pollution from offshore installations, the deliberate dumping of waste from ships and aircraft, and, within certain limits, other forms of pollution from ships. The rights of other states in the EEZ are related to transit and communications, and comprise the freedoms of navigation, overflight by aircraft and the laying of submarine cables and pipelines.

Both Norway and the USSR have claimed EEZs. Norway claimed an EEZ with effect from the beginning of 1977.[31] Although the legislation embodying this claim applies in principle to Svalbard, it has not (yet) been extended to Svalbard, in part at least because the legislation may not be wholly compatible with Norway's international obligations under the 1920 Svalbard Treaty (discussed in Chapter 2). Instead, in June 1977 Norway established a 200-mile fishery protection zone around Svalbard.[32] The USSR did not initially claim an EEZ: instead, like a number of other developed states in the late 1970s, it first claimed a 200-mile fishery zone.[33] Only in 1984 was this replaced by a 200-mile EEZ.[34] As a result of these Norwegian and Soviet claims, the whole of the Barents Sea is now EEZ, with the exception of a triangular-shaped area roughly in the middle of the Barents (see Map 3.2).

Beyond the EEZ lie the high seas. This area is open to use by all states. This principle is usually referred to as the freedom of the high seas. This freedom, which is found both in the 1958 and 1982 Conventions and customary law, includes, but is not limited to, the freedoms of navigation, overflight, fishing, research and the laying of submarine cables and pipelines. In exercising its freedom of the high seas, each state is required to have reasonable regard to the interests of other states in their exercise of the freedom of the high seas. A further feature of the freedom of the high seas is that, subject to a limited number of exceptions, each state has exclusive jurisdiction over its own ships on the high seas. In the Barents Sea the only area of high seas is the relatively small triangular-shaped area referred to above, and even then the sea-bed of this area is continental shelf.

Summing up the jurisdictional zones into which the Barents Sea is divided, around most of its periphery is a band of territorial sea, four miles in width off the Norwegian coasts, twelve miles off the Soviet coasts, measured in most cases from straight baselines. Beyond this the sea-bed is wholly continental shelf. The overlying waters are EEZ, save for a small triangular-shaped area of high seas in the middle of the Barents Sea.

Having now, it is hoped, given adequate background information on the Barents Sea, we can turn to consider the first of the long-standing unresolved legal questions mentioned at the beginning of this chapter, namely the question of whether the 1920 Svalbard Treaty applies to the maritime zones of Svalbard lying seawards of its territorial sea.

2 Application of the Svalbard Treaty to maritime areas

INTRODUCTION

As explained at the end of the previous chapter, the aim of this chapter is to explore the first of the long-standing unresolved legal questions relating to the Barents Sea, namely the question of whether the 1920 Svalbard Treaty (with its regime of equality of treatment for all its parties in the exploitation of the resources of Svalbard) extends to the maritime zones of Svalbard beyond its territorial sea, which (as we saw in the previous chapter) comprise the continental shelf and a 200-mile fishery protection zone. This is a question of considerable practical importance, not only because of the significant resources of those zones but also because exploitation of those resources may have important implications elsewhere in the Barents Sea.

The chapter begins with a brief survey of these resources and their exploitation. It then goes on to explain how the Svalbard Treaty came into being and to give an outline of its provisions. This is followed by a discussion of the legal principles that apply to interpretation of the Treaty. Under the Treaty Norway is obliged to issue regulations to govern mining on Svalbard. The next section of the chapter discusses the legal status of this Mining Code. This is followed by a discussion as to whether the Treaty and Mining Code apply, first in the territorial sea, and, second, in the maritime zones beyond the territorial sea. Whether Norway is entitled to claim such zones in respect of Svalbard is also discussed.

USES OF THE MARITIME AREAS AROUND SVALBARD

As was pointed out in the previous chapter, the maritime areas around Svalbard cover important nursery and fishing grounds. North-east Arctic cod, haddock and capelin migrate between the waters off the Norwegian mainland, Svalbard and the USSR. Historically 25 per cent of the cod catches have been taken in the Svalbard area. As will be discussed in Chapter 4, the migration pattern of these fish stocks necessitates that conservation measures implemented around Svalbard are co-ordinated with corresponding measures taken in the 200-mile zones of the Norwegian and Soviet mainland. The migration area of other fish stocks is restricted to the zones off the Norwegian mainland and Svalbard, whereas shrimps stay exclusively in the Svalbard zone. For all stocks it is, however, important to ensure proper conservation measures to prevent depletion by the large international fishing fleet habitually fishing in this area.

Oil activity on Svalbard started as early as 1918, when the Arctic Oil Company A/S started drilling on land in Grønfjorden. This was, however, without any success.[1] In the beginning of the 1960s there was a certain oil activity, followed by a quiet period until the middle of the 1980s, when interest for oil exploration once again increased. Up to 1985–6 twelve exploration wells had been drilled. There is no great hope of oil discoveries on land, but due to the relatively modest exploration activity and the incomplete data, it is too early to rule out the possibility of discoveries.[2]

Being a part of the Barents Sea, the continental shelf around Svalbard is promising for oil discoveries. It is hard to tell which parts of the Barents Sea are the most promising,[3] but it is claimed that at least that part of the shelf south-east of Svalbard is of particular interest.[4] This area is north-west of the area 'Barents Sea South' designated as part of the Norwegian continental shelf (see Map 5.1). In 1985 an area of shelf up to 74°30′N, which arguably could be part of Svalbard's continental shelf because Bear Island (part of Svalbard) lies between 74°20′ and 74°32′N, was opened by the Norwegian Government for seismic activities.[5] The Norwegian Foreign Minister, Mr Stoltenberg, stated in 1987 that Norway would allow a step-by-step increase of exploration in the Barents Sea including the areas around Svalbard.[6] In 1989 the Barents Sea South area, also extending to 74°30′N, was opened for exploration (with the exception of the Troms II area). The possible problems relating to the application of the Svalbard Treaty to the continental shelf were, however, not addressed either by the Norwegian Government or the Storting (Parliament) when considering the opening of this area for exploration.[7] Although the Barents Sea South has been opened for exploration, oil drilling has so far not commenced in areas which could be regarded as part of the Svalbard shelf.

In addition to the economic interests connected with the maritime areas around Svalbard, these areas are, as pointed out in the previous chapter, of paramount strategic importance. This is partly because of the general strategic importance of the Barents Sea, but particularly because the ice-free waters between Svalbard and the Norwegian mainland are of great importance as a transit passage from the naval bases on the Kola Peninsula to the high seas of the Atlantic Ocean. This means that the security implications of activities and measures implemented in this area will be closely examined among the states concerned. Security concerns may also block measures that would be desirable from a strict resource management perspective.

THE ORIGINS AND AN OUTLINE OF THE SVALBARD TREATY OF 1920

Some account of Svalbard's earlier history was given in the previous chapter. During the height of the whaling era in the seventeenth and eighteenth centuries Svalbard's legal status was disputed. The King of Denmark–Norway claimed sovereignty over Svalbard. The claim was partly based on the assumption that Svalbard was a part of a continuing land mass from the old Norwegian territory of Greenland to Novaya Zemlya, and partly on the claim to sovereignty over the Northern Oceans. Agreements were made between Denmark–Norway and the catching nations, but these agreements should probably not be seen as a common international recognition of Norwegian sovereignty over Svalbard.[8]

The sovereignty claims by Denmark–Norway vanished, however.[9] In the nineteenth century there was a general consensus among interested states that Svalbard was *terra nullius*. But in the years preceding the First World War it had become manifestly obvious that the existing *terra nullius* regime was inadequate to regulate affairs in the archipelago. The discovery of coal in 1899 had led to a number of companies from several different states exploring for and mining coal on Svalbard in competition and often in conflict. There were also conflicts among hunters.

In 1907 Norway asked other interested states to try to negotiate an international arrangement for Svalbard based on the principle of *terra nullius*. The result was a three-power conference between Norway, Sweden and Russia, which accepted the principle of *terra nullius* and suggested establishing a commission consisting of these three nations to govern the archipelago.[10] But when other states subsequently attended the negotiations, it was, in the time before the First World War, impossible to reach agreement on the basis of this three-power model.

The Svalbard question was one of the territorial disputes addressed by the 1919 Peace Conference in Versailles, which resulted in the Svalbard Treaty of 1920.[11] At the Peace Conference the majority view was that the most satisfactory way to regulate activities on the archipelago was to give Norway sovereignty over Svalbard, while maintaining the existing access of other states to the islands for the purpose of carrying on mining, hunting, and other economic activities.[12] The preamble to the Treaty expresses the desire of the parties, while 'recognising the sovereignty of Norway over the Archipelago of Spitsbergen, including Bear Island', as being to see 'these territories provided with an equitable regime, in order to assure their development and peaceful utilisation.'

Under article 1 of the Treaty, the parties 'undertake to recognise, subject to the stipulations of the present Treaty, the full and absolute sovereignty of Norway over the Archipelago of Spitsbergen'. The latter is defined as Bear Island, an isolated island lying roughly half-way between the north Norwegian mainland and the main islands of the archipelago, and all other islands lying between 10° and 35° East and 74° and 81° North. The Norwegian sovereignty over Svalbard recognised in article 1 is, however, subject to a number of far-reaching restrictions.

First, under articles 2 and 3 nationals of all parties to the Treaty 'enjoy equally the rights of fishing and hunting' and may engage in 'all maritime, industrial, mining and commercial operations on a footing of absolute equality'. Second, as regards mining, under article 8 Norway is obliged to provide mining regulations, which require approval from other states parties. Third, the Norwegian authorities' power to tax those who are enjoying the rights given by articles 2 and 3 is restricted. Under article 8 any taxes levied must be devoted exclusively to Svalbard and 'shall not exceed what is required for the object in view'; furthermore, the duty on mineral exports which may be levied is not to exceed one per cent of the value of the minerals exported. Finally, the archipelago is to a great extent demilitarised. Under article 9 'Norway undertakes not to create nor to allow the establishment of any naval base' on Svalbard, and 'not to construct any fortification in the said territories, which may never be used for warlike purposes'.

PRINCIPLES GOVERNING INTERPRETATION OF THE SVALBARD TREATY

The Svalbard Treaty is old and contains few provisions. This means that the Treaty in many cases fails to provide explicit answers to the often complex legal problems of today, including those to be discussed in this chapter. This calls for an examination of the principles by which the Treaty should be interpreted.

Legal theory is generally sceptical of 'rules of interpretation' in international law.[13] The reason is that too much emphasis on such rules may distract the interpreter from what really is the aim of treaty interpretation, namely to find the common intention of the parties. Still, it is necessary both to examine which rules of interpretation the treaty parties have invoked and the support for such rules in legal theory and practice.

Norway argues that those provisions of the Svalbard Treaty imposing obligations on the exercise of Norwegian sovereignty should in principle be interpreted according to a reasonable understanding of their wording. In cases of doubt, however, the solution that should be chosen is that which involves the least limitation on Norwegian sovereignty.[14] This is called the principle of restrictive interpretation.

There is little doubt that Norwegian sovereignty prevails in cases where other states are not granted rights by the Svalbard Treaty. This residual competence follows from the common understanding of sovereignty. But it is not evident that the principle of restrictive interpretation applies when establishing other states' rights.

It is recognised by international courts[15] and accepted by legal theory[16] that a treaty should be interpreted according to the legal situation at the time of its adoption. The question is therefore what the legal status of the restrictive principle was at the time of the adoption of the Svalbard Treaty. The restrictive principle is referred to in inter-war judgments both from arbitral tribunals[17] and from judgments by the Permanent Court of International Justice.[18] This means that this principle was an accepted part of international law at that time.

When examining the judgments of the Permanent Court of International Justice, one finds certain cases where the wording of the relevant treaty explicitly grants other states rights. In such cases the restrictive principle must yield. Thus in the *Wimbledon* case the Permanent Court stated:

> The fact remains that Germany has to submit to an important limitation of the exercise of the sovereign rights which no one disputes that she possesses over the Kiel canal. This fact constitutes a sufficient reason for the restrictive interpretation, in case of doubt, of the clause which produces such a limitation. But the Court feels obliged to stop at the point where the so-called restrictive interpretation would be contrary to the plain terms of the article and would destroy what has been clearly granted.[19]

The *Wimbledon* case also shows that even in such cases the Court will examine the other relevant factors in the case to ensure that the wording really reflects the common understanding of the parties. In this case the

Court discussed the object and purpose of the treaty, the relation of the provision in issue to the other provisions of the treaty, and to similar treaties.[20]

Another situation is where the wording of a treaty indicates that certain rights rest with the state having sovereignty over a territory. Does this mean that other states' rights are excluded? One of the questions decided by the Court in the *Memel* case suggests the opposite solution. Memel was a territory under Lithuanian sovereignty but with a certain autonomy. One of the questions in the case was whether the Lithuanian Governor with the acceptance of the local Directorate had the competence to dissolve the locally elected Chamber. The Statute of Memel, in article 12 paragraph 5, provided: 'The Chamber may be dissolved by the Governor in agreement with the Directorate. The elections to the new Chamber shall take place within six weeks from the dissolution.'[21]

This wording indicates that there was complete freedom of dissolution if the Governor and the Directorate were in agreement. The Court stated, however, that the Directorate must have obtained the confidence of the Chamber before it was entitled to participate in a dissolution. This result was based on the intention of the parties to the Treaty, the relation of article 12(5) to other provisions of the Treaty, and the fact that 'the text of the Statute does not exclude' this solution.[22]

Similarly, the Permanent Court in the *Oder* case explicitly stated that the restrictive principle would not automatically be decisive if the wording of the Treaty was not clear:

> Nor can the Court, on the other hand, accept the Polish Government's contention that, the text being doubtful, the solution should be adopted which imposes the least restriction on the freedom of States. This argument, though sound in itself, must be employed only with the greatest caution. To rely upon it, it is not sufficient that the purely grammatical analysis of a text should not lead to definite results; there are many other methods of interpretation, in particular, reference is properly had to the principles underlying the matter to which the text refers; it will be only when, in spite of all pertinent considerations, the intention of the parties still remains doubtful, that that interpretation should be adopted which is most favorable to the freedom of States.[23]

The general conclusion that may be drawn from these judgments by the Permanent Court is that the restrictive principle will not automatically be decisive whether the wording of a treaty is clear or not: the Court has always analysed the other sources of law closely to find the common intention of the parties to the treaty. This approach was used in the inter-war period and is used today. But, in accordance with the requirement to apply a contemporaneous interpretation, the restrictive principle should, as indicated in the *Oder* case, apply to the interpretation of the Svalbard Treaty 'when, in spite of all pertinent considerations, the intention of the parties still remains doubtful'. In such cases the restrictive principle is applied as an *independent* principle of interpretation. It is, however, usually possible to establish a sufficiently well-founded opinion about the intention of the parties to the treaty when 'all pertinent considerations' are taken into account, and such application of the restrictive principle is therefore not likely to occur very often.

Instead of applying the restrictive principle as an independent principle, a restrictive interpretation will more often be the result of applying other legal sources which are relevant in treaty interpretation. In such cases the restrictive interpretation is not an independent principle, but the *name of a result* reached on the basis of other factors.

The principle of effective treaty interpretation has been considered as the opposite of the principle of restrictive interpretation. The principle of effective treaty interpretation has often been called the principle of 'major purposes' or of 'extensive interpretation'.[24] This means that according to this principle the interpreter should emphasise the major purposes of the treaty, rather than preserving the sovereign freedom of states. The principle of effectiveness in treaty interpretation is also an accepted part of legal doctrine[25] and practice.[26] But this principle cannot be considered as an independent principle in the sense that an extensive interpretation should be used in cases of doubt. An extensive interpretation only applies if justified by other factors of treaty interpretation.

This means that, except for cases where the principle of restrictive interpretation should be applied as an independent principle, the other factors of treaty interpretation will decide whether a restrictive or an extensive interpretation should be applied.[27] Accordingly, a careful examination of all factors relevant in treaty interpretation must be undertaken, instead of automatically applying one or the other principle of interpretation. These factors include *inter alia* the wording of the provision to be interpreted, the general and specific purposes of the treaty, preparatory works, the circumstances surrounding the conclusion of the treaty, the treaty's context and subsequent practice among states parties. Such an examination must be concrete and related to the provision in question. This will be done as regards the question of the application of the Svalbard Treaty in the maritime zones of Svalbard in the following section of this chapter. We find it, however, appropriate to present the 'major purposes' of the Svalbard Treaty at this point, as a general background for the subsequent discussion.

The Svalbard Treaty should be considered to have three principal purposes. The first was to place the Svalbard archipelago under the sovereignty of a single state – Norway – so that the islands would be subject to a proper legal regulation, which the previous *terra nullius* status of the islands was thought incapable of realising. Norway was chosen because of its interests on Svalbard, Svalbard's geographical adjacency to the Norwegian mainland and the need to find a final solution.[28] But the intention was not generally to provide economic benefits to Norway. On the contrary, the Treaty's second purpose was to ensure preservation of the rights which other states had to exploitation of the islands' economic resources according to the previous *terra nullius* status. This was done by ensuring equal access to economic activities (articles 2 and 3) and the requirement that all taxes collected were to be used on Svalbard (article 8 (2)). The third purpose of the Treaty was to secure peaceful development on the islands. Reference is here made to the preamble of the Treaty ('an equitable régime, in order to assure their development and peaceful utilisation') and article 9 of the Treaty (the demilitarisation clause).

One will see that the first and third purposes may well lead to the same result: the peaceful development of the islands may be enhanced by leaving as much control as possible with the Norwegian authorities. This indicates that a restrictive interpretation could be used when it comes to questions regarding the limits of Norwegian control over activities on Svalbard.

If there were indications that the drafters wanted to reserve certain benefits to Norway, a restrictive interpretation could also be applied to limit other states' rights. But the drafters of the Svalbard Treaty placed great emphasis on the need for equal treatment deriving from other states' *terra nullius* economic rights. An extensive interpretation as regards such economic rights would thus generally seem to be in conformity with the drafters' intentions.

Contrary to the typical situation where states obtain rights over another state's territory, Norway did not grant other states' rights over Svalbard, thereby limiting her earlier sovereign rights. Norway was given sovereignty on condition that other states retained their previous extensive *terra nullius* rights. This package deal aspect of the Svalbard Treaty could be used as an argument against restrictive interpretation of the Treaty. It could also be argued that application of the restrictive principle would imply that limitation on other states' rights should be interpreted restrictively, as they had more extensive rights under the *terra nullius* regime than under the Treaty regime. There is, however, no indication in the Svalbard Treaty that Norwegian sovereignty generally was to be of an inferior quality compared to other states' sovereignty over their territory. This means that Norway should not necessarily be barred from claiming a restrictive interpretation of the Treaty. To what extent such an interpretation should be applied in a specific case would then depend on the considerations developed above.

THE LEGAL STATUS OF THE MINING CODE

As mentioned earlier, under article 8 of the Svalbard Treaty Norway is obliged to adopt mining regulations for Svalbard. Such a Mining Code was adopted by Norway by Royal Decree on 7 August 1925.[29] There is, however, disagreement as to whether this Code is legally binding as an international instrument; i.e. is the Code an international agreement (treaty) or is it simply a piece of Norwegian domestic legislation? [30] The answer to this question is important because it determines whether Norway has the right to change the Mining Code without the consent of other states.

If the Mining Code were a treaty, it would have been natural to issue the Code as a document signed by the states parties to the Svalbard Treaty. There are, however, no formal requirements for creating a binding treaty.[31] The decisive factor is to what extent the parties intended to establish a binding treaty. The legal status of the Code must therefore be determined on the basis of an interpretation of the Svalbard Treaty. The Mining Code is not annexed to the Svalbard Treaty as an integral part of the Treaty, unlike the regulations for dealing with claims to the land of the archipelago, which are annexed to the Treaty and 'have the same force and effect as the present Treaty' (article 6). This may also indicate that the Mining Code is not a treaty.

On the other hand, article 8(1) of the Svalbard Treaty states that 'Norway undertakes to provide . . . mining regulations'. It is not said that Norway may *adopt* these regulations. The meaning of the Norwegian competence in this respect is set out in article 8 (4). This provision states that Norway was to provide a draft Mining Code. But if any of the signatory states objected to the draft, the Mining Code would be adopted by an international commission consisting of one representative from each of these states. The commission would take its decisions by a majority vote. If the Mining Code were adopted by such a commission, there would be no doubt that it would be binding for Norway as an international act. This method was not used, because

Norway through negotiations with the other parties managed to reach agreement on the content of the mining regulations. Other states could, however, have used their veto and thereby forced Norway to bring the case before such a commission. Instead, they accepted the draft presented by Norway. The fact that a commission was not necessary should not, however, be decisive for the legal status of the Mining Code. The Mining Code should therefore be regarded as a treaty. This view is supported by article 10 of the Svalbard Treaty, which provides that article 8 was to enter into force when all the signatory powers had ratified the Treaty, whereas the rest of the Treaty was not to enter into force until the Mining Code was in force. The Mining Code was thus seen as a condition for the rest of the Treaty system.

Furthermore, given the procedures governing the adoption of the Mining Code, it would be odd if Norway were free unilaterally to amend the Code, especially since amendment of the Svalbard Treaty would undoubtedly require the consent of the other states parties. We therefore conclude that the Mining Code is an international treaty and that amendments to the Code must follow the same procedure as its adoption.

DOES THE TREATY APPLY IN THE TERRITORIAL SEA?

The various provisions of the Svalbard Treaty do not necessarily all have to have the same geographical application. Article 1 of the Treaty refers to a number of islands by name, as well as to 'all islands great or small and rocks appertaining thereto' within an area defined by certain co-ordinates. Some of the subsequent articles expressly refer to maritime areas, such as 'territorial waters',[32] whereas other articles refer to 'the territories specified in article 1'.[33] The questions that need to be addressed are: first, what effect the area within the Treaty co-ordinates has on the Treaty's geographical application; second, the geographical application of the Treaty provisions expressly referring to maritime areas and; third, the geographical application of those provisions referring only to 'the territories specified in article 1'.

Article 1 states that Norwegian sovereignty is recognised in respect of 'the Archipelago of Spitsbergen, comprising, with Bear Island or Beeren-Eiland, all the islands situated between 10° and 35° longitude East of Greenwich and between 74° and 81° latitude North, especially [certain named islands], together with all islands great or small and rocks appertaining thereto'. The co-ordinates indicated form a box containing islands and maritime areas (see Map 3.2). It could be argued that the 'Archipelago of Spitsbergen' is the whole area within this

Svalbard box, including maritime areas. It is, however, expressly stated in article 1 that the 'Archipelago of Spitsbergen' comprises 'all the islands' within the Svalbard box. The wording of the Treaty therefore clearly indicates that the function of the co-ordinates is only to set out which islands and rocks are to be covered by the Treaty regime, not to make the regime applicable to the maritime areas within the box. This would also be a sensible way of defining the geographical application of the Treaty regime, instead of naming all islands and rocks. Furthermore, there is no reason why the Treaty drafters would want the Treaty to cover exactly the maritime area formed by the co-ordinates. Our conclusion is therefore that the Svalbard box is irrelevant to deciding the extent to which the Treaty applies to maritime areas.[34]

Some of the Treaty articles expressly state that they apply in the territorial sea. Article 2(1) provides that the states parties shall enjoy equally the right to fishing and hunting 'in the territories specified in article 1 and in their territorial waters'. Article 2(2) establishes Norway's right to adopt conservation measures, which shall always apply equally in 'the said regions, and their territorial waters'. Article 3(2) prescribes the principle of equal rights as regards 'maritime, industrial, mining or commercial enterprises both on land and in the territorial waters'.[35] The term *eaux territoriales* was proposed by Norway to replace *eaux adjacentes*. The Norwegian proposal was not founded on any explicit reasons and it was approved by the Spitsbergen Commission without discussion.[36] The application of these provisions in the four-mile territorial sea established by Norway around Svalbard raises no legal problems.

Article 3(1) has, however, a slightly different wording, as it accords the right to 'equal liberty of access and entry' to 'the waters, fjords and ports of the territories specified in article 1; subject to the observance of local laws and regulations, they [i.e. nationals of states parties to the Treaty] may carry on there without impediment all maritime, industrial, mining and commercial operations on a footing of absolute equality'. This paragraph is rather confusing. What do 'the waters, fjords and ports of the territories specified in article 1' or the word 'there' in the second half of this sentence refer to? 'There' could mean either 'the waters, fjords and ports of the territories specified in article 1' or simply 'the territories specified in article 1'. The former seems the more natural because 'the waters, fjords and ports of the territories specified in article 1' seems to be a single phrase.

This brings us to the 'waters, fjords and ports' of 'the territories specified in article 1'. 'Ports' and 'fjords' appear to raise no real difficulties of meaning. 'Waters' is more problematic. This expression should, however, be considered to refer at least to the territorial sea, the reason being that this provision concerns the 'access and entry' to the Treaty area. When the Treaty allows (in article 3(2)) equal rights to certain activities in 'territorial waters', the equal access to carry on those rights must cover at least the same area.

Let us then look at article 1 and the articles referring to 'the territories specified in article 1'.[37] The geographical application of these articles is more difficult to determine. In favour of the view that these articles refer to land territory only are the following arguments. First, article 1 only mentions land areas (various islands). Second, the term 'territory' usually connotes land, and not adjacent waters, unless qualified by some such adjective as 'maritime'. Third, the reference to 'territorial waters', as in the phrase 'the territories specified in article 1 and in their territorial waters', in some of the other articles (such as article 2(1)) would be redundant if 'the territories specified in article 1' included more than land territory.

Against the view that 'the territories specified in article 1' are limited to land territory is the fact that the French text, which is equally authentic, does not use the term *territoires* or something similar, but speaks of *les régions visées à l'article 1er*. This is legally much less precise than 'territories', and is not necessarily restricted to land. Furthermore, there is a close link between land territory and sea territory in international law. The 1958 Convention on the Territorial Sea and the Contiguous Zone[38] in article 1(1) provides that '[t]he sovereignty of a state extends, beyond its land territory and its internal waters, to a belt of sea adjacent to its coast, described as the territorial sea'. This seems also to have been the position at the time before the Svalbard Treaty's adoption. Thus, in the Grisbådarna case of 1909, the tribunal stated:

> Considérant que cette opinion est conforme aux principes fondamentaux du droit des gens, tant ancien que moderne, d'après lesquels le territoire maritime est une dépendance nécessaire d'un territoire terrestre.[39]

This 'necessary connection' between land territory and sea territory indicates that the point of departure should be that a treaty applying to land should also apply to the territorial sea. Neither of the two authentic texts of the Svalbard Treaty excludes the application of all the Treaty's provisions in the territorial sea. The key point after this would be to consider to what extent an antithetical interpretation of the Treaty's provisions referring to the territorial sea necessitates the conclusion that that the remaining provisions are limited to applying to land territory only. A closer examination of the Treaty provisions referring to 'the territories specified in article 1' is necessary.

These articles are diverse in character. Under article 3 (3) ships have the right to call at Norwegian ports; article 4 deals with public wireless telegraph stations; article 5 with an international meteorological station; article 7(1) with the ownership of property; article 8(1) with Norway's duty to provide mining regulations; article 8(2) with tax regulations; article 9 is concerned with the prohibition against the establishment of naval bases and fortifications and the use of the territory for warlike purposes; while, finally, article 10(2) deals with claims from Russian nationals and companies.

Limiting the geographical application of most of these articles to land territory is a perfectly possible solution and does not create special difficulties. It would, however, be a very artificial solution if articles 2 and 3 applied in the territorial sea whereas article 8(2) on tax limitations did not so apply. The reason is that the purpose behind the equal economic rights and the tax limitations is the same: Norway should not gain special economic benefits from her sovereignty over Svalbard.

The danger to peaceful development of the area is the same whether fortifications are made on the land territory of Svalbard or in its territorial sea. It is hard to imagine that the drafters of the Svalbard Treaty would accept fortifications in the territorial sea around Svalbard when such installations on land were not accepted. Article 9 should therefore also apply in the territorial sea. The *travaux préparatoires* show that Norway had proposed that article 9 should apply *dans les ports et eaux territoriales*, but this reference was deleted by the Spitsbergen Commission on the basis of a proposal from the United Kingdom. There are no recorded reasons for this amendment and any change of substance should thus not be presumed.[40] This is both an argument in favour of the application of article 9 in the territorial sea and an indication that the distinction between application on land and in the territorial sea was not considered by the Spitsbergen Commission.

It would thus seem that the reason for not mentioning territorial waters in articles 8(2) and 9 is that they concern activities that were not regarded as maritime when the Treaty was drafted. By contrast, articles 2 and 3, which include typical maritime activities such as fishing and navigation, are made applicable to territorial waters. Such a background to the different wording of the two sets of provisions makes an antithetical interpretation of articles 2 and 3 by no means compelling.

On the contrary, we would argue, on the basis of the provisions of article 9 and especially of article 8(2), that application of these articles in the territorial sea would be the only solution conforming to the system for allocation of rights in the Svalbard Treaty. This would also conform to the general rule that a treaty applying to land territory should also be considered to apply to sea territory, i.e. the territorial sea. Articles 8(2) and 9 should therefore apply in the territorial sea. It would be a possible solution that articles 8(2) and 9 applied in the territorial sea, whereas the other articles not referring to the territorial sea should apply only to the land territory of Svalbard. There is, however, no reason to accept a different geographical application of the articles which all refer to article 1. We also find that the arguments in favour of applying all articles of the Treaty in the territorial sea are so convincing that there is no room for a restrictive interpretation using the restrictive principle as an independent principle. Consequently, all the provisions of the Svalbard Treaty apply in the territorial sea.

DOES THE MINING CODE APPLY IN THE TERRITORIAL SEA?

The geographical application of the Mining Code also needs examination. Section 1 of the Mining Code has similar wording to article 1 of the Treaty, and thus a similar question arises as to what extent the Mining Code applies in the territorial sea.

Article 3(2) of the Svalbard Treaty provides that states parties have equal rights to mining in the territorial sea. As already mentioned, article 8(1) establishes that Norway is to provide, in accordance with certain procedures, a Mining Code 'for the territories specified in article 1'. We have concluded that this expression includes the territorial sea around Svalbard. Norway is accordingly under a duty to provide mining regulations for the territorial sea. The question is then whether the existing Mining Code applies to the territorial sea or whether Norway has to provide regulations specifically for the territorial sea.

We have argued that the point of departure should be that a treaty applying to land territory should also be considered to apply to the territorial sea. Consequently, the Mining Code should apply to the territorial sea. Article 1 of the Svalbard Treaty and section 1 of the Mining Code contain similar wording regarding their geographical application, which also indicates that the Code applies in the territorial sea. In the following we will discuss to what extent other factors imply an application limited to the land territory.

Certain provisions of the Mining Code make it possible to argue that its application is limited to land areas. Section 7(1) provides that the search for minerals 'may be made on one's own property as well as on that of any other party, and on the Public Lands'. This indicates that the Mining Code only applies on land under private or governmental ownership.

The system for obtaining mining licences is also a factor in favour of the Code's application only on land. Section 9 (1) states that anyone who has discovered a mineral deposit acquires a right to the discovery provided he, in the presence of two witnesses, 'by marks in solid rock or, by other lasting and satisfactory means, visibly locates a discovery point'. This may be difficult to do at sea. It is of course possible to mark the spot by a buoy or similar object, but this does not have the 'lasting' effect as marks 'in solid rock'.

Section 9(2)(d) states that a notice of discovery shall contain 'a sample, handed over at the same time, of the minerals or rocks found'. As regards oil exploration, it has been accepted by the parties to the Treaty that 'geological indications' of oil may be presented instead of a sample.[41] The reason for this is that due to the large investments necessary for oil exploration, explorers need a guarantee before such investments arc made. This method could, however, also be used in maritime areas.

The Mining Code is based on the principles that the first finder obtains the mining rights, that there are provisions on the geographical size of claims, and that specific work duties are imposed in order for the finder to keep his claim. These principles may not be very suitable for application at sea, but on the other hand we do not think their application at sea would raise insurmountable difficulties.

The need for Norwegian control over any petroleum activity could also be invoked as a factor in favour of a restrictive interpretation of the Code, implying that it does not apply in the territorial sea. There is, for example, a vital need to protect workers and the environment against the dangers of petroleum activity. But this need also exists for petroleum activity on the land areas of the islands, and the Mining Code has never been seen as a restriction on Norwegian competence to establish stringent safety regulations. The safety of workers and the environment may therefore be ensured even if the Code applies in the territorial sea.[42]

It has been argued above that one of the main purposes of the Svalbard Treaty was to preserve the equal rights to economic activities on the islands. This purpose is fulfilled by the Treaty itself, as we have considered it to apply in the territorial sea. There is not then the same need also to apply the Mining Code. Neither is there a need to regulate the relationship between the land owner and the mining claimant at sea. But, on the other hand, there is the same need for a system for allocation and regulation of mining rights at sea as on land. In this respect the sea-bed under the territorial sea is very similar to the situation of the 'Public Lands' on Svalbard (cf. the Mining Code section 7(1) already discussed).

The system and the principles of the Mining Code point in the same direction, namely that the Code is not applicable in the territorial sea. Since the Code has been drafted with mining on land in mind, application of the Code in the territorial sea would also raise difficult questions of interpretation. On the other hand, there are no indications that the drafters of the Mining Code ever considered the geographical application of the Code or deliberately chose to limit it to land territory. Furthermore, Norway always has the possibility of changing the Code by a new internationally agreed interpretation according to today's needs – similar to what was done when 'geological indications' were accepted as the basis for acquiring mineral rights. If a change through interpretation proves too difficult, the ultimate possibility will be to have the Code changed by the procedure envisaged in the Treaty's article 8. On balance, we therefore think that there is not sufficient reason to deviate from our point of departure, namely that since the Mining Code applies on the land areas of Svalbard, it should also be considered to apply in the territorial sea and that the Svalbard Treaty and the Mining Code should have the same geographical application.

Applying the Code in the territorial sea is also supported by subsequent state practice. The Norwegian Ministry of Industry has in a letter to the Svalbard Commissioner of Mines stated that the Mining Code does apply in the territorial sea,[43] which has also become the practice of the Norwegian government on Svalbard.[44] This must be seen as a Norwegian interpretation of the Code and no states have protested. This state practice carries a certain weight, supporting the Mining Code's application in the territorial sea. So far, only seismic activity has been undertaken in the territorial sea; there is no experience of the practical application of the Mining Code to full-scale oil exploration or exploitation. Nevertheless, this limited practice indicates that the Mining Code may be applied in the territorial sea without too much difficulty.

IS NORWAY ENTITLED TO CLAIM A CONTINENTAL SHELF AND A 200-MILE ZONE IN RESPECT OF SVALBARD?

Before discussing whether the Svalbard Treaty applies to maritime zones beyond the territorial sea, an important question is whether there is anything, either in the Treaty itself or in the general body of international law, which prohibits the generation of such maritime zones in respect of Svalbard. First, as regards the Treaty: the competence of a state to claim maritime zones beyond the territorial sea derives from its sovereignty over that territory. Since Norway has sovereignty over Svalbard (article 1 of the Treaty), and since there is nothing in the Treaty which either expressly or by implication restricts Norway's competence to claim maritime zones in respect of Svalbard (and the Treaty in fact, as we have seen, expressly refers to Svalbard's territorial sea), it must follow that Norway has the competence to claim maritime zones around Svalbard.[45]

Second, is there anything in the general body of international law which would prohibit the generation of maritime zones around the islands of the Svalbard archipelago? Under the pre-UN Convention on the Law of the Sea (1982) law all islands were entitled to a territorial sea,[46] continental shelf,[47] and exclusive fishing zone.[48] This position is followed in principle by the 1982 Convention, apart from referring to the Exclusive Economic Zone (EEZ) instead of the exclusive fishing zone. A limited exception is, however, made in article 121(3), for 'rocks which cannot sustain human habitation or economic life of their own': such islands cannot generate a continental shelf or EEZ. Two questions arise from this provision: first, is it applicable law today; and second, does it apply to any of the islands of the Svalbard archipelago? As regards the first question, the 1982 Convention has not yet come into force. Therefore the only basis on which article 121(3) can be applicable law today is if it has already passed into existing customary international law. In view of current state practice this seems most unlikely.[49] Second, if the 1982 Convention entered into force and were ratified by Norway, would article 121(3) apply to any of the islands of the Svalbard archipelago? There are undoubtedly many islets and rocks in the archipelago to which article 121(3) would apply, but they lie so close to the main islands of the group (which clearly are much too large to be considered as 'rocks') that in practice the generation of the continental shelf and EEZ would not be affected. The only islands lying at any distance from the main archipelago which might possibly be covered by article 121(3) are Bear Island and Hope Island. Both have a few people living on them all the year round, principally scientists and meteorologists, but the latter are so dependent on the outside world that it is questionable whether the islands can be said to 'sustain human habitation or economic life of their own'. As to whether the islands are 'rocks', there is no definition of 'rock' in the 1982 Convention, but with an area of 178 square kilometres (Bear Island) and 47 square kilometres (Hope Island), the islands must be considered as being larger than any reasonable definition of a 'rock'. It thus seems that article 121(3) would have no practical application to the Svalbard archipelago.

The general conclusion that follows from this discussion is that there is nothing in either the Svalbard Treaty or the general body of international law which prevents Norway from claiming the whole range of generally accepted maritime zones around Svalbard. Except for the Soviet Union, which rejects the legal validity of the 200-mile fishery protection zone which Norway has claimed around Svalbard,[50] Norway's competence to claim maritime zones around Svalbard does not appear to be questioned by any state.

DOES THE TREATY APPLY BEYOND THE TERRITORIAL SEA?

Norway argues that other states cannot claim any rights under the Svalbard Treaty on the continental shelf and in the 200-mile zone around Svalbard. First, this position is based on the wording of the Treaty, which is considered to apply in the territorial sea only. Norway claims that although the drafters did not foresee the possibility of jurisdiction outside the territorial sea, this cannot change the Treaty's geographical application. The reason is that the drafters were fully aware of the restrictive interpretation of treaties, and that Norwegian sovereignty over Svalbard would be general and would always apply when the Treaty does not contain specific limitations.[51]

Second, it is argued that there is a continuous continental shelf extending northwards from the mainland coast of Norway up to and beyond the Svalbard archipelago. Norway has therefore sovereign rights to this shelf independently of the Svalbard Treaty.[52] Fleischer argues that the provisions in international law governing delimitation only apply where there is a continuous shelf between two states. As Svalbard is not another state in relation to Norway, there is no specific Svalbard shelf.[53]

It is true that Norway would have ordinary continental shelf rights on the shelf surrounding Svalbard if there were no archipelago of Svalbard. Geologically, there is a continuous continental shelf stretching northwards from northern Norway to Svalbard and beyond: the deepest point between northern Norway and Svalbard is about 450 m. The whole of this sea-bed is therefore legally continental shelf. The question is, however, what effect the archipelago of Svalbard has on the continental shelf. Is there a Svalbard shelf which is governed by the Svalbard Treaty?

The Norwegian line of thought can obviously not justify Norwegian sovereign rights in the 200-mile zone around Svalbard, as Svalbard is more than 200 miles away from the Norwegian mainland. The 200-mile zone around Svalbard must therefore be based on Norwegian sovereignty over Svalbard, and, accordingly, on the Svalbard Treaty.

The exclusive economic zone (EEZ) concept is an integration of sovereign rights to the continental shelf and the waters within 200 miles.

When sovereignty over Svalbard is being used to establish a 200-mile zone, this is an argument in favour of considering the rights to the continental shelf to stem from sovereignty over Svalbard, not from the sovereignty over northern Norway. Similarly, the fact that Norway uses Svalbard as the basis for a median line delimitation of the continental shelf towards the USSR and Greenland, indicates that the continental shelf around Svalbard is a Svalbard shelf, not a mainland shelf.

But other factors also support the view that the continental shelf around Svalbard appertains to Svalbard rather than northern Norway. First, as concluded above, Svalbard has the ability under international law to generate a continental shelf. This follows from Norwegian sovereignty over Svalbard. Moreover, this right to the shelf exists *ipso facto* and *ab initio*. This means that Svalbard has a continental shelf automatically, without any positive proclamation by Norway.

Second, if Svalbard were a separate state or belonged to a state other than Norway, it would have its own continental shelf and a delimitation of the continental shelf between Norway and that state would be necessary. Now, Norway has both sovereignty over the mainland and Svalbard, and, as stated by Fleischer, Svalbard is not a separate state. But Svalbard has a distinct legal regime different from that of the Norwegian mainland. It is therefore natural that the continental shelf generated by Svalbard is governed by this special regime and that a delimitation towards the Norwegian mainland shelf is necessary. Such a delimitation has been effected between the Svalbard 200-mile zone and the mainland zone (see below, p. 44). Our conclusion is therefore that Svalbard has its own continental shelf.[54] The next question, then, is, whether other states may claim their special Treaty rights on the Svalbard continental shelf and in the 200-mile zone.

As with the question of the application of the Svalbard Treaty in the territorial sea, the geographical application of each of the Treaty's articles must in principle be discussed separately. In the following, articles 2 and 3 will be discussed first, both because these are the only articles referring to the territorial sea and because they concern traditional maritime activities. These articles, it will be recalled, provide that the nationals of other parties enjoy equal rights of *inter alia* 'fishing and hunting' and 'mining' in Svalbard's 'territorial waters'.

There does not appear to have been a uniform view on the legal nature of the territorial sea in 1920. Some states (including the nations of the British Empire, USA and the Scandinavian states) regarded the territorial sea as a single belt of waters over which a coastal state had sovereignty, while other states (such as France, Italy and Russia) regarded it as a zone or zones (often of varying widths) where a coastal state had jurisdictional competence for specific purposes.[55] The former view, as is well known, in due course became the accepted conventional and customary law. The latter view, however, is not incompatible with a coastal state's rights to sea-bed and fishery resources beyond four miles (the present width of Svalbard's territorial sea), even though under present-day international law such rights are juridically different from the territorial sea. But let us in the following assume that the former view is the correct interpretation of the phrase 'territorial waters' in the Svalbard Treaty.

Norway is right in claiming that other states are not granted any rights outside the territorial sea according to the wording of the Svalbard Treaty. A literal interpretation of the Treaty would therefore indicate that Norwegian sovereignty should prevail beyond the territorial sea. This is the point of departure for the discussion that follows. Norway also relies on the principle of restrictive interpretation of treaties when claiming that other states cannot claim Treaty rights on the continental shelf and in Svalbard's 200-mile zone. The application of this principle and other factors relevant to interpretation of the Treaty will now be discussed.

It is of course idle to speculate what the intentions of the drafters of the Treaty were in 1920, since they did not know about maritime resource zones beyond the territorial sea. On the one hand, one could argue that if other states had known about maritime zones beyond the territorial sea, they would have applied articles 2 and 3 of the Treaty to them. On the other hand, we have no guarantee that Norway would have accepted such extensive rights for other states on the continental shelf and in the 200-mile zone.[56]

At the time the Treaty was concluded, its provisions on equal rights in relation to fishing, mining and other activities set out in articles 2 and 3, extended, however, not only to the land territory of Svalbard, but also to the full extent of the only maritime resource zone then known to international law, the territorial sea. A plausible argument would be that since the maritime zones now known to international law also include the continental shelf, fishing zone and EEZ, the Treaty should be extended to them, to the extent that such zones have been claimed in respect of Svalbard.

This supposition follows from the fact that the application of articles 2 and 3 to the territorial sea was uncontroversial when the Treaty was being drafted (and is found in the draft Treaty which the Norwegian Government presented to the Spitsbergen Commission)[57] and from the idea of equal rights which permeates the Treaty. One of the principal purposes of the Svalbard Treaty therefore indicates that the rights of other states should be extended by analogy. A preliminary conclusion

of measures for regulating fisheries, particularly quotas. The application of the non-discrimination principle for fisheries in the Svalbard 200-mile zone will be discussed in Chapter 4. Suffice it to say here that the extension of the Treaty to zones beyond the territorial sea would not introduce fundamentally new legal problems. The question of how the Treaty's non-discrimination principle applies to different commercial activities, and especially to the regulation of fishing and hunting in Svalbard's internal waters and territorial sea, has long existed. Thus, although the extension of the Treaty would raise legal problems relating to fishery regulations, these are not qualitatively different from existing problems relating to the interpretation and application of the Treaty.

The next set of problems which could arise if the Svalbard Treaty applied beyond the territorial sea relates to exploration and exploitation of hydrocarbons on the continental shelf. First, the present Mining Code is not well suited to sea-bed mining (as explained earlier). It is, however, a perfectly possible solution that the Treaty applies beyond the territorial sea, whereas the Mining Code does not. There is accordingly no need to discuss the geographical application of the Mining Code in connection with the Treaty's geographical application, and the Code will therefore be discussed separately below.

There is no prohibition in the Treaty against Norway adopting environmental and safety regulations for future oil activity on the Svalbard shelf.[64] On the contrary, article 2 of the Treaty provides for the enactment of proper conservation measures. Such measures have been adopted for mining activities on the land territory of Svalbard. It could be argued that the application of other states' equal mining rights under the Treaty to the Svalbard shelf would lead to a too extensive oil activity, to the detriment of both environmental and strategic concerns. But Norway must be entitled to limit the activity for environmental reasons. In our view, Norway must also have the competence, or even the duty, to take into consideration strategic factors when regulating the extent of oil activity. Reference is here made to the preamble and article 9 of the Treaty, which clearly indicate the importance the drafters attached to the peaceful utilisation of Svalbard.

A preliminary conclusion would be that the extension of the Svalbard Treaty to maritime zones beyond the territorial sea would raise some legal problems which application of the Treaty confined to the land territory of Svalbard does not raise. These problems are, however, not very numerous or of great complexity. Nevertheless, the existence of these problems would support an argument that the Svalbard Treaty should be restrictively interpreted.

But, as we have already concluded, articles 2 and 3 apply to all maritime zones known at the time of the Svalbard Treaty's adoption. The intention to preserve equal rights to economic activities according to the former *terra nullius* regime was an essential purpose of the Treaty. This desire is also formulated in provisions throughout the Treaty and forms an important part of the context of articles 2 and 3. Hence, strong support is rendered to an analogous application of articles 2 and 3 on Svalbard's continental shelf and in the EEZ.

We turn now to the guidance which may be derived from international courts. The International Court of Justice has several times stressed the close nexus between land territory and appurtenant maritime zones: 'it is the land which confers upon the coastal state a right to the waters off its coasts';[65] 'the rights of the coastal state in respect of the area of continental shelf that constitutes a natural prolongation of its land territory into and under the sea exist *ipso facto* and *ab initio*, by virtue of its sovereignty over the land, and as an extension of it';[66] 'the land is the legal source of the power which a state may exercise over territorial extensions to seaward';[67] 'continental shelf rights are legally both an emanation from and an automatic adjunct of the territorial sovereignty of the coastal state. It follows that the rights of a coastal state comprise, *ipso iure*, the rights of exploration and exploitation over the continental shelf to which it is entitled under international law';[68] 'the capacity to engender continental shelf rights derives ... from sovereignty over the land mass'.[69] This close link between land territory and maritime zones suggests that, because Norwegian sovereignty over Svalbard is subject to limitations, and since Norway's right to maritime zones beyond the territorial sea derives from her sovereignty over Svalbard, it should follow that Norway's rights in the maritime zones around Svalbard are also subject to the same limitations. An extension of other states' rights by analogy to the continental shelf and the 200-mile zone is thus supported.

Three cases need more detailed study, however, as they involve factual situations which may be considered close to the problem of the application of the Svalbard Treaty on the continental shelf: namely *Petroleum Development Ltd v Sheikh of Abu Dhabi*,[70] *Petroleum Development (Qatar) Ltd v Ruler of Qatar*[71] and *Aegean Sea Continental Shelf Case*.[72] In the *Abu Dhabi* case the Umpire, Lord Asquith, had to decide whether a concession granted by the Sheikh of Abu Dhabi to Petroleum Development Ltd, whereby the Sheikh purported to transfer the exclusive right to drill for oil in the 'whole of the lands which belong to the rule of the Ruler of Abu Dhabi and its dependencies and all the islands and the sea waters which belong to that area', extended to the continental shelf. Lord Asquith held that the concession did not so extend. In 1939, when the concession was granted, the continental shelf doctrine had not been mooted. Nor, in Lord Asquith's opinion, had the continental shelf yet (in 1951) become settled international law. Thus, in Lord Asquith's view, 'it would be a most artificial refinement to read back into the contract the implication of a doctrine not mooted till seven years later, and ... not even today admitted to the canons of International Law'.[73] The facts of the *Qatar* case were similar, and the arbitrator, Lord Radcliffe, reached the same decision as Lord Asquith. However, his decision, unlike that of Lord Asquith, contains no reasons.

At first sight these two cases seem strong precedents in support of the Norwegian government's view. However, there are a number of factors which weaken their value. First, the law applied in the two cases was not international law, but 'principles rooted in the good sense and common practice of the generality of civilized nations – a sort of "modern law of nature" ' (the *Abu Dhabi* case),[74] and 'the legal principles familiar to civilised nations' (the *Qatar* case). Second, the two cases involved interpretation of a concession. The Svalbard Treaty is not of the nature of a concession. Norway has not granted rights to the other parties: instead the Treaty is a bargain whereby Norway obtained sovereignty and the other parties various rights in relation to Svalbard. As concluded above, when interpreting the Svalbard Treaty, it is necessary both to take into account the need for effective Norwegian control, but also the purpose of preserving the *terra nullius* equal economic rights. It is thus inappropriate only to apply to the Svalbard Treaty the restrictive principle of interpretation which was applied to the concessions at issue in the *Abu Dhabi* and *Qatar* cases. Furthermore, it is explicitly stated in the *Abu Dhabi* case that Lord Asquith did not use 'the rule that grants by a sovereign are not to be construed against the grantee'.[75] Finally, the fact that Lord Asquith found that there was no such thing in law as the continental shelf, could lead one to argue that his finding that the concession did not extend to the continental shelf was essentially *obiter*. However, it seems fairly clear from Lord Asquith's judgment that he would have found that the concession did not extend to the continental shelf, even if the latter had existed in international law. Our conclusion is, therefore, that these two cases give very limited support to the Norwegian legal stance.

Support for an extension of the Treaty to the continental shelf by analogy may be derived from the judgment of the International Court of Justice in the *Aegean Sea Continental Shelf* case. In this case Greece sought to found the Court's jurisdiction *inter alia* on the General Act for Pacific Settlement of Disputes. Greece had acceded to the General Act with a reservation excluding 'disputes relating to the territorial status of

Greece'. Greece argued that this reservation did not include the continental shelf. One of its arguments was that the continental shelf was unknown in international law in 1928 when the General Act was signed and in 1931 when Greece acceded. The Court rejected the Greek argument. It held that the 'territorial status of Greece' was a generic term denoting any matters comprised within the concept of territorial status under general international law, and there was thus a presumption that its meaning evolved with international law. It therefore followed that 'disputes relating to the territorial status of Greece' included disputes relating to the continental shelf.

Two observations must be made in assessing the value of this case as a precedent for application of the Svalbard Treaty. First, the latter does not use a broad generic term such as 'territorial status' found in the Greek reservation to the General Act. But, on the other hand, articles 2 and 3 of the Treaty include all the jurisdictional areas known at the Treaty's adoption in 1920, including the territorial sea, indicating an extensive interpretation. Second, different considerations may apply to the Svalbard Treaty than to jurisdictional clauses relating to the International Court. As the Court commented in relation to Greece's attempt to rely on the *Abu Dhabi* arbitration: 'there is an essential difference between a grant of rights of exploration and exploitation over a specified area in a concession and the wording of a reservation to a treaty by which a state excludes from compulsory procedures of pacific settlement disputes relating to its territorial status.'[76] The basic purpose of ensuring continued equal economic rights, however, makes the Svalbard Treaty different from the grant of a concession in the *Abu Dhabi* case. The justification for using an analogy from the *Aegean Sea Continental Shelf* case is more convincing.

The time has come to reach a conclusion on the geographical application of articles 2 and 3. The wording of these articles does not provide a clear solution, and, consequently, using the principle of restrictive interpretation as an *independent* principle could be relevant. On the basis of decisions by the Permanent Court of International Justice, we have, however, concluded that the restrictive principle should only be applied in such a way if other factors do not offer sufficient guidance to form a conclusion. The factors discussed above should, however, provide ample basis for a conclusion.

We have found that the wording of the Svalbard Treaty cannot be regarded as an expression of the intention of the parties to limit other states' rights under articles 2 and 3 to the territorial sea. These two articles were made applicable to all maritime zones known at the time of the Treaty's adoption. The general purpose of preserving the former *terra nullius* economic rights of other parties to the Treaty indicates that articles 2 and 3 should be applied to the continental shelf and the 200-mile zone by analogy. Besides, anomalies would be created if articles 2 and 3 applied within but not outside the territorial sea. Finally, the close link between sovereignty over land areas and sovereign rights to maritime zones beyond the territorial sea indicates that limitations to sovereignty over land should also apply to sovereign rights in maritime zones. The problems connected with the application of articles 2 and 3 beyond the territorial sea are on the other hand not of such great complexity that they should prevent such application. This leads us to the conclusion that articles 2 and 3 should, by analogy, be regarded as applicable on the continental shelf and in the 200-mile zone of Svalbard.[77]

But what about the other articles of the Treaty? The condition for recognising other states' Treaty rights on the continental shelf and in the 200-mile zone of Svalbard is, of course, that Norway has jurisdiction over the relevant activities under international law: in relation to other activities other states' rights will be based on the freedom of the high seas. The jurisdiction of the coastal state in these maritime areas is functionally limited, the principal content of the jurisdiction being sovereign rights to exploration and exploitation of the natural resources. This means that several of the articles in the Svalbard Treaty are without any interest beyond the territorial sea, for example article 4 on rights as regards public wireless telegraph stations. We will concentrate our discussion on article 8(2) (on taxation) and article 9 (on fortification and military use of the archipelago).

In contrast to articles 2 and 3, the geographical application of articles 8 and 9 cannot be extended by analogy by reference to their explicit application in the territorial sea. Application of a restrictive interpretation would thus seem more appropriate. But, on the other hand, there are also here certain factors offering assistance in interpretation.

The close link between land territory and the maritime zones beyond the territorial sea is relevant also when examining the geographical application of articles 8 and 9. We concluded above that all the articles of the Treaty apply in the territorial sea. The same considerations are relevant when discussing the application of these articles beyond the territorial sea. It must, however, be examined to what extent such an application creates special problems.

Article 8(2) of the Svalbard Treaty provides that any taxes levied on Svalbard must be devoted exclusively to Svalbard. This would result in offshore activities being taxed at a much lower rate than on the continental shelf of the Norwegian mainland or almost any other continental shelf. But this limitation on the right of taxation is scarcely a legal problem, but a consequence that is undesirable from the point of view of the Norwegian treasury (there is, however, no prohibition against higher taxation by a company's state of registry): furthermore, the position would be the same if oil were discovered onshore in Svalbard. Finally, the equal economic rights and the tax limitation are closely linked: Norway was not to derive special economic benefits from Svalbard. Consequently, on the basis of the same line of reasoning as developed for its application in the territorial sea, article 8(2) should apply in the maritime zones beyond the territorial sea.

The only aspects of article 9 that would seem capable of application to maritime areas are the provisions not to use Svalbard 'for warlike purposes' or to construct 'any fortifications' therein. This would raise legal questions: first, as to what is meant by 'fortification' (would it cover the emplacement of weapons or submarine listening devices on the sea-bed?);[78] and, second, what is meant by 'warlike purposes' (would these cover transit of the waters by naval vessels – which seems most unlikely – or the use of the waters for naval manoeuvres or weapons testing?). It is enough here to show that the application of article 9 to areas beyond the territorial sea of Svalbard would raise legal problems, without attempting to suggest what the solution to these problems is.[79] Our conclusion is that the problems with applying article 9 beyond the territorial sea are of not much greater complexity than the corresponding problems connected with its application on land. Furthermore, there is the same need to promote the peaceful utilisation of the areas beyond as the areas within the territorial sea. Thus, on the same basis as its application in the territorial sea, article 9 should therefore also apply on the continental shelf and in the 200-mile zone. Similarly, we would assume that the application of the other articles in the Svalbard Treaty in these areas would not create such great difficulty that their application should be denied.

Pursuant to its view on the maritime zones around Svalbard, Norway has made its general continental shelf legislation applicable beyond the territorial sea of Svalbard. The Continental Shelf Law of 1963 was considered to be applicable on the shelf around Svalbard,[80] and the Petroleum Law of 1985, which replaced the 1963 Act, has the same application.[81]

Other states have, however, either protested against the Norwegian view or reserved their position. The Soviet Union is among the states protesting.[82] The United Kingdom claims that the Svalbard Treaty applies to the Svalbard shelf.[83] Only Finland has supported Norway.[84] So far, no exploration or exploitation on Svalbard's continental shelf has started and there has not been an immediate need to find a solution. But as oil drilling goes on in other parts of the Barents Sea and the prospects for hydrocarbons on the Svalbard shelf are positive, there is a need to solve the question of the legal status of the Svalbard shelf.

It has been necessary to regulate fishing activities off Svalbard. To avoid conflicts with other states, Norway established in 1977 a non-discriminatory 200-mile fishery protection zone. Norway has, however, established this zone pursuant to the Norwegian Law on the economic zone of 1976, and has stated that Norway has the right to establish a discriminatory EEZ around Svalbard. All states, except the Soviet Union, recognise Norway's right to establish a 200-mile zone. But these other states do not accept the Norwegian view that the Treaty only applies within the territorial sea. They have either protested or reserved their position also in this respect.[85] Although Norway has claimed that all regulations adopted in the Svalbard 200-mile zone for fishery protection are non-discriminatory, there have been episodes in 1986, 1987 and 1988 involving foreign fishing vessels challenging Norwegian jurisdiction by serious violations of the fishery regulations. The Norwegian Coast Guard inspects fishing vessels in the fishery protection zone, but so far violations have only resulted in warnings, not arrests. The problems connected with fisheries management in the 200-mile zone around Svalbard are further discussed in Chapter 4.

Existing state practice does not provide any guidance as to the Svalbard Treaty's application beyond the territorial sea. Oil drilling has not commenced on Svalbard's shelf, and the states involved still stick to their original positions. It could be argued that the Norwegian policy of non-discrimination and the lack of arrests during the thirteen years since the fishery protection zone was established should be seen as acceptance by Norway that the Treaty applies in this zone and that Norway's enforcement jurisdiction is limited. Norway has, however, always claimed that her policy was a result of the need for low tension in this region, not of the acceptance of other states' legal position. The need for avoiding conflicts is of obvious importance and Norway's policy in this respect should not undermine her legal position. But if a long period of time elapses without any success in practice for the Norwegian position, this position will be weakened.

DOES THE MINING CODE APPLY BEYOND THE TERRITORIAL SEA?

We have concluded that article 8(1) of the Svalbard Treaty applies on Svalbard's continental shelf. This means that Norway is obliged to provide mining regulations for the shelf. As with the territorial sea, the question is whether the existing Mining Code applies to the continental shelf or whether Norway has to provide new regulations.

We would argue that the existing Mining Code applies on the continental shelf of Svalbard. Despite certain factors pointing to the opposite solution, we concluded above that the Code applies in the territorial sea. This was mainly because a treaty applying to the land territory should usually also be considered to apply to the territorial sea and because the Svalbard Treaty and the Mining Code should have the same geographical application. The last argument is equally sound for the Code's application on the continental shelf and we have also demonstrated the close relationship between a coastal state's rights to the land territory and the concomitant rights to the continental shelf.

Contrary to the situation in the territorial sea, however, there is no state practice to support the Mining Code's application on the shelf. Furthermore, it will be even more difficult to apply the Mining Code on the continental shelf than in the territorial sea. It is, for example, hard to imagine how the principle of the first finder's right should be applied on the continental shelf. The Norwegian government would probably not be required to open an area for oil drilling until it considered it justifiable on the basis of relevant technical and environmental concerns, rather than because a claimant provided positive 'geological indications' for oil discoveries. But after an area had been opened for exploration the wording of the Code would give preference to the first finder. Such a system would not promote an orderly development of continental shelf activity.

In such cases it is necessary to interpret the Code in a way that would, to the greatest possible extent, respect the assumed intention of the parties to the Treaty: in particular, the right to equal access must be secured. Similar problems connected with application of the Mining Code to new needs have, however, arisen in connection with oil drilling on land. 'Geological indications' have, for example, been accepted instead of delivering a sample of the relevant mineral when applying for a claim patent. Adjustments have also been made through interpretation when marking a discovery point for minerals in the territorial sea. Such adjustments are also possible beyond the territorial sea. If it becomes too difficult to reach a practicable solution through interpretation, the Code may be amended according to the procedure laid down in article 8 of the Treaty. It should therefore be possible, either through interpretation or amendment, to apply the Mining Code also on the continental shelf of Svalbard, and thereby respect the object and purpose of the Treaty and the Mining Code.

CONCLUSIONS

We have concluded, on the basis of a legal analysis, that the Svalbard Treaty and the Mining Code apply in Svalbard's territorial sea and in the maritime zones beyond the territorial sea. We are, however, of the opinion that from a political point of view a solution based on full Norwegian sovereignty (i.e. non-application of the Treaty in these zones) is to be preferred. The reason is that full sovereignty would prevent conflicts over the interpretation of the Svalbard Treaty and the Mining Code and would make it easier to implement effective regulatory measures and enforce them. On the other hand, we have also concluded that such difficulties are not of such complexity that application of the Svalbard regime in maritime areas should legally be precluded.

Oil exploration on the Svalbard shelf is only in its initial phase and drilling has not yet been started. It appears that the dispute over the Svalbard Treaty's geographical application may be preventing the commencement of drilling. Pressure to allow drilling may, however, mount if commercial discoveries are made in other parts of the Barents Sea. In the meantime, environmental and strategic interests will benefit from minimal oil activity on the Svalbard shelf.

Fishing has, however, been going on all the time around Svalbard, which makes fisheries management necessary. Norway has established a 200-mile fishery protection zone around Svalbard and introduced conservation measures. Conflicts regarding the application of the Svalbard Treaty in the 200-mile zone may, however, prevent an effective management regime. This will be discussed further in Chapter 4.

Norway must in the near future assess the need for a solution to the question of whether the Svalbard Treaty's regime applies on the continental shelf and in the 200-mile zone around Svalbard. The likeliest such solution might well be that Norway accepted the Treaty regime's application in these areas. On the other hand, the possibility of achieving the most desirable solution (i.e. that the Treaty regime does not apply), must also be considered. We will undertake a concluding examination of these questions in Chapter 6.

3 Boundary delimitation in the Barents Sea

INTRODUCTION

In 1957 Norway and the USSR signed an agreement which establishes a maritime boundary between them in the Varangerfjord.[1] Paragraph 1 of article 1 of the agreement provides that 'the sea frontier' between Norway and the USSR in the Varangerfjord 'shall follow a straight line' from the terminus of their land frontier to 'the intersection of the outer limits of Norwegian and Soviet territorial waters'. The outer limit of Norway's territorial sea (waters) is four miles, measured from a straight baseline drawn between the initial point of the boundary line and Cape Kibergnes; the outer limit of the USSR's territorial sea is twelve miles, measured at the time the agreement was signed from the low-water mark. Paragraph 2 of article 1 provides that 'neither of the Contracting Parties shall extend its territorial waters beyond the straight line' running from the intersection referred to above to the mid-point of a line drawn across the mouth of the Varangerfjord between Cape Kibergnes (in Norway) and Cape Nemetskiy (in the USSR), this point being some 13.5 nautical miles (hereafter all references to miles are to nautical miles) from the two Capes, the nearest land (see Map 3.1). The two boundary lines described in paragraphs 1 and 2 are 12.6 and 11.75 miles in length respectively.

The line described in paragraph 1 of article 1 is a territorial sea boundary proper. The line described in paragraph 2 would become a territorial sea boundary if either party were to extend its territorial sea beyond the breadths of four miles and twelve miles that they had in 1957. In fact neither party has done so. However, paragraph 2 became relevant in 1985 when the USSR changed its baseline in the Varangerfjord from the low-water mark to a straight baseline. The effect of paragraph 2 was to prevent the USSR's territorial sea extending beyond the line described in that paragraph, which it otherwise would have done in some places.

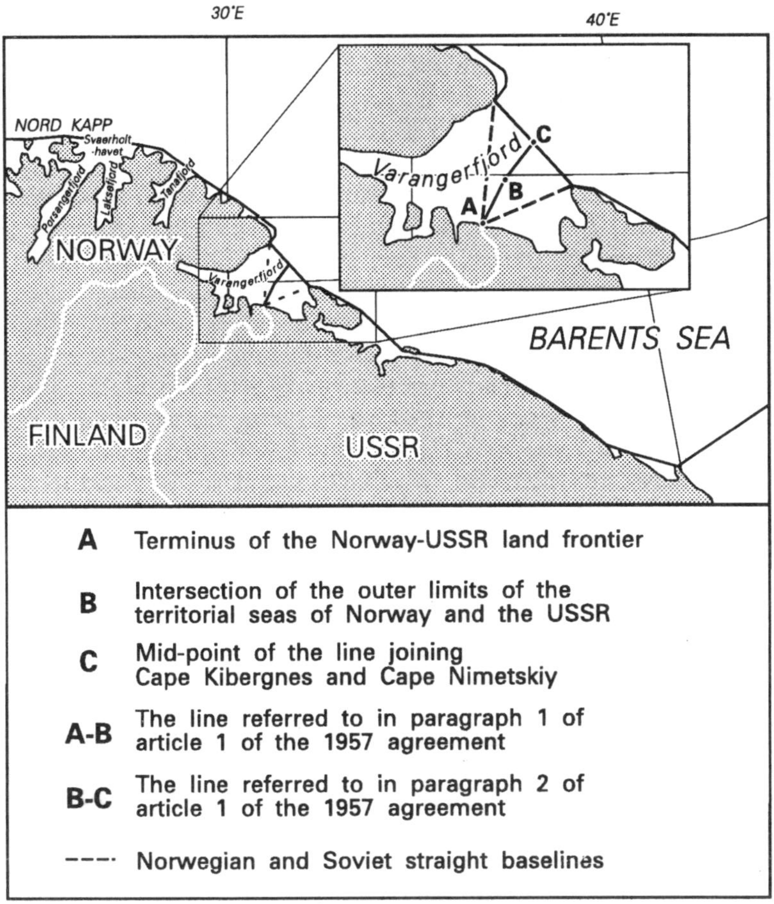

Map 3.1 Norway–USSR maritime boundary in the Varangerfjord

At the time the agreement was signed in 1957, neither Norway nor the USSR possessed any other maritime zones. Since 1957, however, as we saw in Chapter 1, each has claimed a continental shelf and an exclusive economic zone (EEZ). Although paragraph 2 of article 1 of the 1957 agreement is limited in its wording to the line it describes being a potential boundary for the territorial sea only, it would be contrary to the spirit of the agreement for either party to claim that its continental shelf or EEZ extended beyond that line. Therefore, the line may in practice, though not in strict law, also be regarded as a continental shelf and EEZ boundary.

The boundary just described obviously covers only a tiny part of the Barents Sea, indeed it hardly reaches out into the Barents Sea proper. In the vast remainder of the Barents Sea, no boundary between the maritime zones (continental shelf and economic zone) of Norway and the Soviet Union has yet been agreed. The absence of such a boundary is an obstacle both to the effective management of living resources (as will be seen in Chapter 4) and to the exploration and exploitation of offshore oil and gas (as will be seen in Chapter 5), as well as being a source of tension in itself, as explained later in this chapter.

This chapter examines the origins of the dispute between Norway and the USSR over a maritime zones boundary in the Barents Sea and the negotiations that have so far taken place to try to resolve this dispute. It subjects each party's position to critical scrutiny, and offers a possible solution. The final part of the chapter considers the question of whether it is possible to explore for or exploit oil and gas in the disputed boundary area before a boundary has been agreed. First, however, the chapter begins by giving an outline of the rules of international law governing maritime boundary delimitation.

AN OUTLINE OF THE INTERNATIONAL LAW OF MARITIME BOUNDARY DELIMITATION

For the sake of readers who are not familiar with the rules of international law governing the delimitation of maritime boundaries, it seems desirable to begin this chapter by giving a brief and succinct account of the rules relating to the drawing of boundaries between the maritime zones with which we are concerned in this chapter (the continental shelf and economic zone), before examining the dispute over the absence of such boundaries in the Barents Sea. Readers who are familiar with the relevant law will appreciate the difficulty, if not impossibility, of giving a concise account of the rules that is very meaningful.

As with all areas of international law, there are both rules found in treaties and rules found in customary international law, which in the case of maritime boundary delimitation are not necessarily the same (as they are in some other areas of international law). We will begin first by looking at treaty rules. In the case of the continental shelf, rules on the drawing of boundaries between neighbouring states' continental shelves are found in two treaties, the Convention on the Continental Shelf, 1958[2] and the United Nations Convention on the Law of the Sea, 1982.[3]

Both Norway and the USSR are parties to the 1958 Convention. Neither is a party to the 1982 Convention, although both have signed it: if and when they both ratify it and the Convention enters into force (which it is still some way off doing), the Convention will prevail over the 1958 Convention as between Norway and the USSR.[4]

Article 6 of the 1958 Convention lays down what is essentially a three-point formula for determining the boundary between the overlapping continental shelves of neighbouring states. It provides that such a boundary is to be 'determined by agreement' between the states concerned.

> In the absence of agreement, and unless another boundary line is justified by special circumstances, the boundary is the median line, every point of which is equidistant from the nearest points of the baselines from which the breadth of the territorial sea of each state is measured.

In other words, states shall (1) first seek to agree on the boundary; (2) if they are not able to do so, the boundary is the median line unless (3) another line is justified by special circumstances. By 'special circumstances' the drafters of the Convention appear to have had in mind such matters as exceptional configurations of the coastline, islands and the presence of navigable channels.

The 1982 Convention takes a rather different approach from the 1958 Convention. Article 83 provides that:

> The delimitation of the continental shelf between states with opposite or adjacent coasts shall be effected by agreement on the basis of international law, as referred to in article 38 of the Statute of the International Court of Justice, in order to achieve an equitable solution.

The reason for this bland, and apparently almost meaningless, provision is that at the UN Law of the Sea Conference there was a deep division of opinion between one group of states, which thought that continental shelf delimitation should be based on equidistance, and another group of states, which wanted delimitation to be based on equitable principles (following the approach of the International Court of Justice, discussed below). Article 83 is the best the Conference could come up with by way of a compromise.

The only treaty provision dealing with delimitation of neighbouring states' overlapping economic zones is article 74 of the 1982 Convention. At the Law of the Sea Conference there appears to have been a feeling that, in general, it is desirable for continental shelf and economic zone boundaries to coincide, and as a result article 74 is the same, *mutatis mutandis*, as article 83 quoted above.

We turn now from the rules on maritime boundary delimitation found in treaties to those found in customary international law. The latter rules are primarily the product of the International Court of Justice and arbitral tribunals. The International Court has dealt with this matter in four cases. These are the *North Sea Continental Shelf* cases[5] (1969), which concerned the continental shelf boundaries between Denmark and West Germany and between the Netherlands and West Germany; the *Continental Shelf (Tunisia/Libya)* case[6] (1982), which concerned the continental shelf boundary between Tunisia and Libya; the *Gulf of Maine* case[7] (1984), which concerned the boundary between the continental shelves and 200-mile fishing and economic zones of Canada and the USA in the Gulf of Maine; and the *Continental Shelf (Libya/Malta)* case[8] (1985), which concerned the continental shelf boundary between Libya and Malta. The leading arbitrations are the *Anglo-French Continental Shelf* arbitration[9] (1977), which concerned the continental shelf boundary between France and the United Kingdom in the English Channel and Western Approaches; and the *Guinea/Guinea Bissau Maritime Boundary* arbitration[10] (1985), which concerned the boundaries between the territorial seas, economic zones and continental shelves of Guinea and Guinea Bissau.[11] At the time of writing two further maritime boundary cases were pending before the International Court (between El Salvador and Honduras, and between Denmark and Norway concerning the boundary between Greenland and Jan Mayen), and there was at least one pending arbitration (between Canada and France over the maritime boundaries between Canada and the French islands of St Pierre and Miquelon).

Although there is now, by international law standards, a considerable body of case law on maritime boundary delimitation, it is very difficult to say with any precision what the rules of customary international law on this subject are. This is so for a number of reasons, as explained by Weil.[12] First, the International Court and arbitration tribunals have not adopted the conventional approach to ascertaining rules of customary international law, which is to examine the practice of states and deduce a customary rule if there is a sufficient degree of practice coupled with evidence that states believe that what they are doing is permitted or required by international law *(opinio iuris)*: instead the courts have simply declared the law. In short, we are dealing here with judge-made law. Second, in making this law, the judges have not been consistent between cases or even within cases – there has been no linear development from case to case. Third, the rules enunciated by the courts are lacking in precision. The courts have not yet succeeded in laying down rules of general application with a sufficient degree of precision; nor of explaining convincingly how the result in a particular case has been derived from the general principles adopted by the courts. For these reasons it is especially difficult to try to present a synthesised account of the courts' case law, especially in the limited space available here.[13]

In the *North Sea Continental Shelf* cases the International Court, after observing that there was no single method of delimitation of which the use was compulsory, stated that 'delimitation is to be effected by agreement in accordance with equitable principles, and taking account of all the relevant circumstances'.[14] This dictum has been reiterated in all subsequent cases: the most recent cases have stressed, in addition, that the goal of the delimitation process is an equitable solution. The courts have, however, never indicated with any precision what the content of such equitable principles is (although they have stressed that the application of such principles is a matter of law and not simply of discretionary justice – or a decision *ex aequo et bono*, to use the language of article 38(2) of the Court's Statute). The courts have, however, indicated various factors (or criteria) which are to be taken into account in applying equitable principles. (In fact in some cases there seems to be a certain blurring between equitable principles and the criteria of delimitation.[15]) Although in the *North Sea* cases the Court suggested that there was no limit to the kind of factors or criteria that might be taken into account in effecting an equitable delimitation, subsequent cases have tended to narrow such factors or criteria to those that are primarily geographical. They include the configuration of the relevant coastlines (for example in the *North Sea* cases, where the concavity of the German coastline coupled with the convexity of the Danish and Dutch coastline meant selection of a method of delimitation other than that of equidistance, which would have produced an inequitable result); obtaining a reasonable degree of proportionality between the length of coastlines of the states concerned and the area resulting to each from the delimitation; in general not allowing islands to have the same effect in determining the boundary as the mainland; and, finally, not allowing the boundary line to encroach on or cut off areas that more naturally belong to one of the states concerned rather than the other(s).

On the other hand, courts have discounted as relevant factors economic differences between the states concerned; differences in the area of land territory belonging to the states concerned; and the natural resources and ecology of the delimitation area (except perhaps in exceptional circumstances). In the *North Sea* cases the International Court, with its repeated references to the continental shelf being a natural prolongation of a state's land territory, placed considerable emphasis on geological and geomorphological factors: the later cases, however, have so downgraded these factors that they are now all but irrelevant – at least for areas within 200 miles of the coast where title to the continental shelf, under the 1982 Convention, is now based on distance.

It will be recalled that in article 6 of the 1958 Convention on the Continental Shelf equidistance is not only specifically mentioned as a method of delimitation but is also given a certain primacy. The courts have been consistently adamant, however, that article 6 has not become part of customary international law and that equidistance is not a method of delimitation whose use is mandatory under customary international law. Nevertheless, the courts have acknowledged that in certain circumstances, particularly where the coasts of the states concerned are opposite one another (as opposed to being adjacent), equidistance can be a useful method of delimitation, certainly as a starting point. In some cases, particularly the *Tunisia/Libya* case, the courts have suggested that prior conduct of the parties, for example agreement on a provisional line of delimitation, may be a relevant factor and could even in some circumstances amount to an estoppel: in other words, the parties would be precluded from claiming areas beyond such a provisional line as part of their maritime zones.

Although it was suggested above that the treaty rules on maritime delimitation (particularly article 6 of the 1958 Convention) differ from the customary rules as elaborated by international courts and tribunals, it should perhaps be pointed out here that a different view was taken by the arbitral tribunal in the *Anglo-French Continental Shelf* case. The tribunal suggested that the equidistance/special circumstances provision of article 6 (which the tribunal saw as a single rule rather than, as suggested above, a two-step process) and the rules of customary law had the same object – the delimitation of the boundary in accordance with equitable principles. The tribunal regarded the rules of customary law as a relevant means both for interpreting and completing the provisions of article 6.

To sum up the international law on maritime boundary delimitation very briefly: the primary rule, both in conventional and customary law, is that the states concerned must make a genuine effort to negotiate and reach an agreement on a boundary. Obviously it is open to the states concerned to reach agreement on any boundary they wish. The significance of the equidistance/special circumstances rule of the 1958 Convention and the equitable principles and criteria of customary law for negotiated boundaries is that those rules and principles which are considered particularly apposite will be invoked by a state to support its negotiating position. Depending on the outcome of negotiations, such rules and principles may or may not be reflected in the maritime boundary agreed. In practice, of about 400 potential maritime boundaries in the world, some 120 have so far been agreed through negotiations by the states concerned. If the states concerned are unable to agree on a boundary, then they may – if they both agree to do so – refer the dispute to some form of third party settlement – the International Court of Justice, arbitration or conciliation. A court or tribunal seized of a boundary dispute will seek to resolve it by applying the rules contained in any treaties to which all/both the disputing states are parties or, if there are no treaty rules applicable, the equitable principles and criteria of customary law, described above. It will be obvious from what has been said that the generality and lack of precision of such principles and criteria, coupled with the fact that every area where a boundary is to be drawn has a unique set of geographical and other characteristics, mean that no precise boundary line will be indicated to the court or tribunal at the outset: the court or tribunal will have to select those principles and criteria which it considers will produce an equitable result. Sometimes this has been done as a single process: in other cases a two-stage process is involved where the court or tribunal first selects a provisional boundary (often the line of equidistance) and then checks, and if necessary corrects, it to ensure that it represents an equitable solution. In practice about a dozen maritime boundary disputes have so far been referred to third party settlement.

ORIGINS OF THE MARITIME BOUNDARY DISPUTE IN THE BARENTS SEA AND THE COURSE OF NEGOTIATIONS SO FAR

Having briefly examined the general rules of international law governing maritime boundary delimitation, we must now turn to consider the particular maritime boundary dispute in the Barents Sea, beginning with its origins, and then going on to see how the rules examined above might be applied to resolve the dispute. As explained at the beginning of this chapter, no boundary has yet been agreed either between the continental shelves of Norway and the USSR or between their economic zones. We will look first at the origins of the dispute over the continental shelf boundary, and then consider subsequently the origins of the dispute over the economic zone boundary.

As we saw in Chapter 1, the legal definition of the continental shelf is rather different from a geographical or geological definition; nor has the legal definition remained constant over time. To recapitulate briefly what was said in Chapter 1: under article 1 of the 1958 Continental Shelf

Convention (to which, as pointed out earlier, both Norway and the USSR are parties) the continental shelf is defined as the sea-bed out to 200 m or beyond if the depth of the superjacent waters admits of the exploitation of the resources of the sea-bed. Under article 76(1) of the 1982 UN Convention on the Law of the Sea the continental shelf is defined as the sea-bed out to 200 miles or the edge of the continental margin, whichever is the further. Although the 1982 Convention is not yet in force, there is a good deal of evidence to suggest that the definition of the continental shelf in the Convention is in the process of passing into customary law:[16] in fact part of the definition, that the continental shelf extends at least to 200 miles, has been held by the International Court of Justice in the *Libya/Malta* case already to have passed into customary international law.[17] Thus under the 1958 Convention at least part of the bed of the Barents Sea is continental shelf – that part lying in less than 200 m of water, together with however much beyond that is exploitable. Under the UN Convention and emerging customary law there is no doubt that the whole of the bed of the Barents Sea is, legally speaking, continental shelf.

Since a state has in law a continental shelf automatically and does not need to make an express claim to one,[18] this means that since the emergence of the continental shelf as a concept in international law (i.e. that a state has sovereign and exclusive rights to explore and exploit the resources of its continental shelf, as defined above) in the 1950s, there has been the need to establish a boundary between the continental shelves of Norway and the USSR. This need was originally probably only of a rather theoretical character: it was not really perceived as a practical issue until each state made a formal claim to a continental shelf. In the case of Norway, this occurred in 1963 when a Royal Decree was issued which claimed sovereign rights over:

> The seabed and the subsoil in the submarine areas outside the coast of Norway . . . as far as the depth of the superjacent waters admits of exploitation of natural resources . . . but not beyond the median line in relation to other states.[19]

The USSR did not make its own claim until five years later, when the Supreme Soviet issued a decree claiming sovereign rights over:

> The seabed and subsoil of the submarine areas adjacent to the coast or to the islands of the USSR . . . to a depth of 200 metres or, beyond that limit, to where the depth of the superjacent waters admits of the exploitation of the natural resources of the said areas

[i.e. the 1958 Continental Shelf Convention formula].[20] The Decree went on to repeat the 1958 Convention's provisions as regards boundaries with neighbouring states (i.e. such boundaries should be determined by agreement with such states: in the absence of agreement, and unless another boundary was justified by special circumstances, the boundary should be the median line).

The first contacts between Norway and the USSR over a continental shelf boundary in fact took place the year before the USSR made its continental shelf claim, in 1967, when Norway proposed that negotiations over a continental shelf boundary should be held. This led to an informal meeting in Oslo in 1970, and the first formal negotiations in Moscow in 1974. Subsequently a further eight rounds of negotiations have so far taken place, at varying intervals.[21] In these negotiations the parties have recognised that the point of departure is article 6 of the 1958 Continental Shelf Convention (referred to earlier). Norway has argued that there are no special circumstances and that the boundary should therefore be the median line between the Norwegian and Soviet mainland coasts and between the Svalbard archipelago and the Soviet archipelagos of Novaya Zemlya and Franz Josef Land. The USSR, on the other hand, has argued that there are special circumstances and that the boundary should be a sector line (i.e. a line of longitude) running from the terminus of the existing boundary in Varangerfjord (described at the beginning of this chapter; see Map 3.1) towards the North Pole, but modified in an easterly direction in the Svalbard area so as to avoid cutting through the area defined in article 1 of the 1920 Svalbard Treaty (referred to in the previous chapter; see Map 3.2). These arguments will be examined in more detail in the next section. The sea-bed lying between the median line proposed by Norway and the sector line proposed by the USSR is an enormous area of approximately 45,000 square nautical miles (155,000 square kilometres) – an area greater than the Norwegian sector in the North Sea – and comprising roughly 11 per cent of the entire Barents Sea. During the negotiations that have so far taken place there does not appear to have been a great deal of change in each side's position, although the Norwegian Government has made it clear that it would be prepared to modify its position on the median line in return for some concessions on the USSR's sector claim. So far the USSR does not appear to have indicated any real willingness to make such concessions, apart from its initiative in 1988 (discussed below).

In 1977 negotiations over a continental shelf boundary became further complicated with the establishment by both Norway and the USSR of 200-mile economic zones.[22] This development meant that future negotiations would be concerned not just with a continental shelf boundary

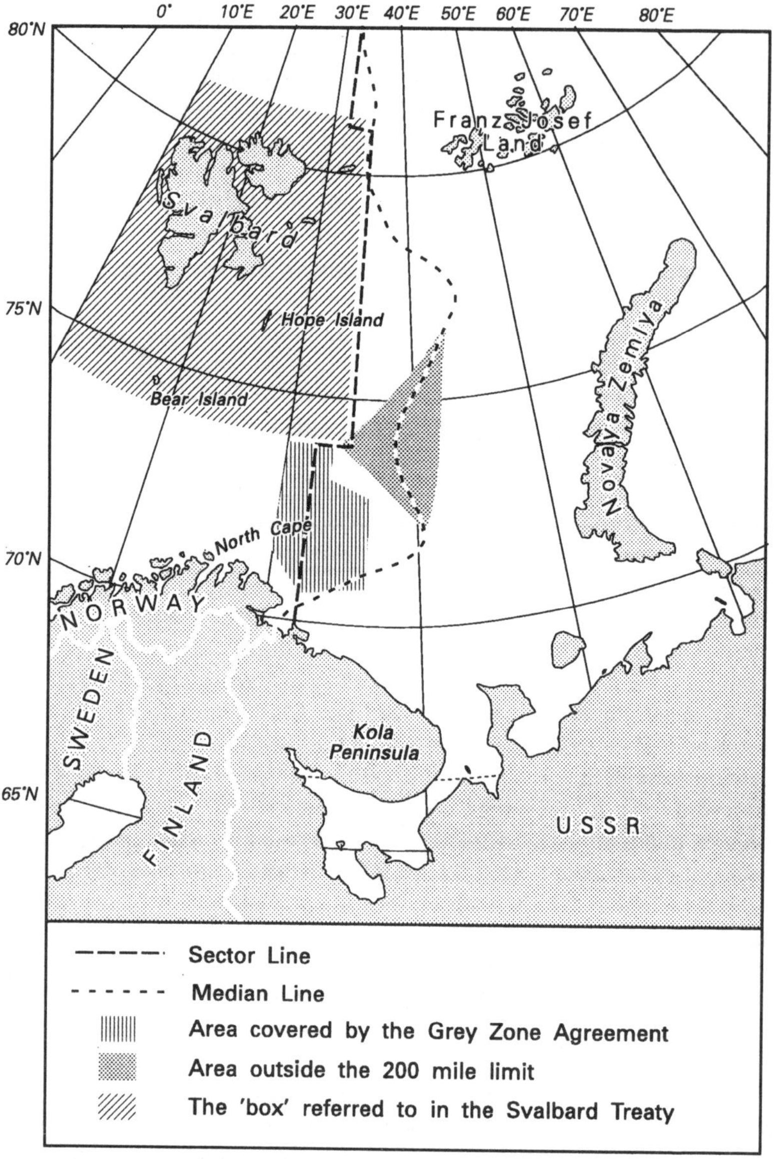

Map 3.2 The median and sector lines in the Barents Sea

but also with an economic zone boundary. In the late 1970s there was no immediate urgency for a continental shelf boundary, as neither Norway nor the USSR desired to begin early exploration for hydrocarbons in the disputed area. The same was not true for an economic zone boundary, however, as the Southern Barents has been long and heavily fished by both Norwegian and Soviet fishermen, and, to a lesser degree, by fishermen from third states. Given the difficulties already apparent by this time in negotiating a continental shelf boundary, and the fact that the boundaries for the economic zone and continental shelf would probably coincide,[23] it was unlikely that Norway and the USSR would rapidly agree on an economic zone boundary. Both sides were agreed, however, on the need speedily to come to some temporary practical arrangement for the exercise of each state's fisheries jurisdiction, particularly in respect of vessels from third states, in the waters lying over the disputed area of continental shelf and economic zone (i.e. the area between the median line in the east and the sector line in the west and within 200 miles of the mainland). Without some such arrangement there would have been unregulated fishing, with a real danger of over-fishing and a resulting threat to the well-being of fish stocks, as well as the likelihood of confrontations and disputes if one state purported to exercise its jurisdiction over the vessels of other states or third states. Negotiations to try to come to some form of arrangement were held in the Spring and Summer of 1977 and led to the signature of the so-called 'Grey Zone Agreement' on 11 January 1978.[24]

The Agreement applies to an area which not only covers a large part of the disputed area of continental shelf and economic zone in the Southern Barents Sea (the grey zone proper), but also to areas to the west of the sector line and east of the median line respectively (see Map 3.3). These areas of undisputed economic zone covered by the Agreement differ considerably in size, however. Whereas the area of undisputed economic zone belonging to Norway covered by the Agreement (area 2 on the map) amounts to 23,000 square kilometres, that belonging to the USSR (area 3) amounts to a mere 3,000 square kilometres. The reason why the area of application of the Agreement is not limited to the area between the median line and the sector line is because the USSR could not accept a solution which would indicate that the latter area was the subject of the boundary dispute between Norway and the USSR.[25] This was in spite of the fact that the Agreement contains a 'without prejudice' clause, which reads:

> [This Agreement] does not prejudice the positions or views of either party with regard to the boundaries of the parties' areas of fisheries jurisdiction, or to the delimitation of the continental shelf and sea areas referred to in this [Agreement].[26]

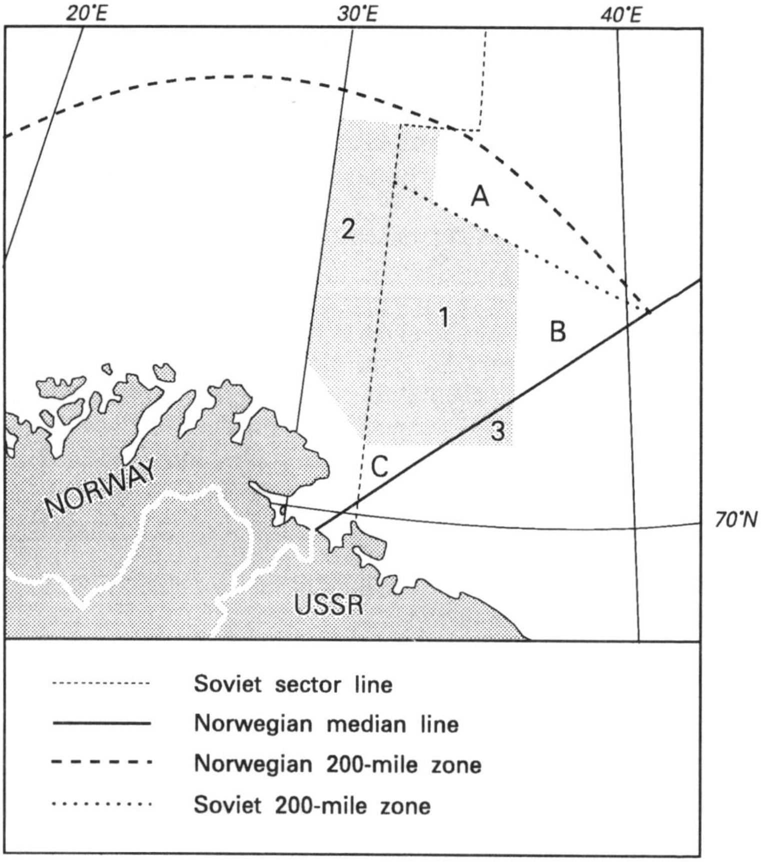

Map 3.3 Area covered by the 'Grey Zone Agreement'

One further point of interest about the area of application of the Agreement should also be noted. The area bounded by the median line, the sector line and a line 200 miles from the Norwegian and Soviet mainland comprises 60,700 square kilometres. Of this area, only 41,500 square kilometres is covered by the Agreement. In other words, there is a considerable area of the economic zones of Norway and the USSR between the sector line and the median line which is not covered by the Agreement and where to a considerable extent the zones overlap. Of this area the Agreement says nothing. The Norwegian Government has explained that the parties assume that that part of this area within 200 miles of the Soviet coast (Map 3.3, areas B and C) is regarded by the USSR as a Soviet area of fisheries jurisdiction as long as the Agreement lasts. As regards that part of this area beyond 200 miles of the Soviet coast but within 200 miles of the Norwegian coast (area A), the parties assume that this area will be high seas as long as the Agreement lasts (though the Norwegian Government does not explain why such an apparently odd assumption has been made). On the other hand, the area within 200 miles of the Norwegian coast north of the area of the Agreement is regarded by Norway as being part of Norway's economic zone.[27]

Of the area of application of the Agreement as a whole, Mr Per Tresselt, a former head of the Department of Legal Affairs in the Norwegian Foreign Office and a member of the Norwegian delegation that negotiated the Agreement, has explained that it was 'a compromise which *inter alia* had regard to important fishing banks in the boundary area'.[28] Nevertheless, the fact that the Agreement covers a much larger area of undisputed Norwegian economic zone than Soviet zone led in Norway to considerable opposition to the Agreement when it was signed: and the Agreement was approved by the Storting (the Norwegian Parliament) by a majority of only four votes.

The substantive provisions of the Agreement are largely taken up with arrangements relating to fisheries. These provisions are discussed in detail in the next chapter. One feature of these provisions should, however, be mentioned here, and that relates to jurisdiction. The Agreement provides that each party is to exercise jurisdiction only in respect of its own fishing vessels and not in respect of vessels of the other party: jurisdiction over the fishing vessels of third states is to be exercised by whichever party has licensed such vessels. This arrangement, which can be characterised as a regime of split jurisdiction, is very important for Norway, as it avoids any suggestion that the grey zone is an area of joint jurisdiction or a kind of condominium. It has been the policy of successive Norwegian Governments since at least 1944 to avoid any kind of arrangements with the Soviet Union in the North which smack of a condominium, because of fears that this would undermine Norwegian sovereignty (particularly over Svalbard) and possibly detach it from the Western alliance.

The Agreement is a temporary one, and was originally concluded for one year only. It has subsequently been extended for annual periods in every year following its signature, and this is likely to continue until agreement is reached on a continental shelf/economic zone boundary. In general the Agreement has worked well, and it is some measure of the Agreement's perceived usefulness and success that its annual renewals have continued uninterrupted, notwithstanding the fact that the non-socialist parties in Norway, who were originally opposed to the Agreement, subsequently replaced the Labour Party as the Government and the fact that a prominent member of the Norwegian delegation that negotiated the Agreement, Arne Treholt, was in 1985 convicted of being a Soviet spy.

As mentioned above, negotiations over a continental shelf (and economic zone) boundary have taken place at intervals, without there apparently having been much change in either party's position. In January 1988, however, came a dramatic new development from the Soviet side, a product no doubt of the effect Mr Gorbachev and his policies of *glasnost* and *perestroika* have had on Soviet diplomacy generally, and more specifically a follow-up to Mr Gorbachev's speech in Murmansk in October 1987, when he called for peaceful co-operation in exploiting the resources of the Arctic.[29] In January 1988 the then Soviet Prime Minister, Mr Ryzkhov, visited Oslo. He suggested that the main reason why negotiations over a boundary appeared to be deadlocked was because the differing interests of Norway and the USSR in the area had become entangled: drawing a line on a map would not achieve any balance between their interests and would not strengthen security and stability. He proposed that instead Norway and the USSR should co-operate in the joint development of oil and gas resources in a 'special economic zone', which would cover both parts of the disputed area and undisputed areas of both parties. Such joint development would be on the basis of full equality and in the form of a joint venture. This proposal was rejected by the Norwegian Government both on grounds of principle, because of its long-standing policy, referred to above, of avoiding any arrangements with the Soviet Union in Northern Waters which resemble a condominium, and on practical grounds, because existing joint development zones in other parts of the world have not worked well. It was agreed that negotiations over a continental shelf and economic zone boundary should continue.[30]

A year later came a further initiative from the USSR. In January 1989 it became known that a Soviet official had secretly visited Oslo the previous month and had told the Norwegian Government that the USSR was willing to accept a modified version of the sector line as the boundary in the northernmost part of the disputed area. At the same time he reiterated the USSR's proposal for a zone of joint co-operation, stressing that its details were open to negotiation.[31] The significance of this episode is that it appears to be the first time in fifteen years of negotiations that the USSR has made any concession on the sector line as the boundary. Subsequently a number of informal meetings between Norwegian and Soviet officials have been held, and there is now a cautious Norwegian hope that the USSR will before long be willing to discuss a compromise boundary.

Having examined the origins of the dispute over the continental shelf and economic zone boundaries and the course of negotiations so far, we will now turn to examine critically the arguments of the parties in the light of the applicable international law.

A CRITICAL EXAMINATION OF THE PARTIES' VIEWS AS TO WHERE THE BOUNDARY SHOULD LIE

This section will evaluate the arguments of Norway and the USSR as to where the boundary between their overlapping continental shelves and economic zones should be drawn – the Norwegian claim that the boundary should be the median line, the Soviet claim that it should be a sector line – in the light of the applicable international law. It must be stressed at the outset, as was pointed out above when discussing the general international law rules relating to maritime boundaries, that international law does not dictate any particular line as the boundary or any particular method of delimitation to be used in determining the boundary. It is open to Norway and the USSR to agree on any particular boundary line or boundary solution they wish. This might therefore suggest that the exercise to be undertaken in this section is rather fruitless. We would suggest that this is not so. As explained above, while states remain free to agree on any solution they wish, they do in practice when negotiating a boundary invoke the various rules and principles of international law already outlined (see pp. 56–61). There is evidence that this has happened with Norway and the USSR in their negotiations over a boundary in the Barents Sea.[32]

At the outset it is desirable to recognise that the disputed boundary has three distinct segments (as the parties also appear to recognise).[33] The first segment runs from the terminus of the existing maritime boundary described at the beginning of this chapter, at the mouth of Varangerfjord, to a point 200 miles from the mainland of either Norway or the USSR (or possibly a point 200 miles from both). In this segment there is a need for both a continental shelf and economic zone boundary between the adjacent mainland coasts of Norway and the USSR. The second segment comprises the area in the middle of the Barents Sea which is more that 200 miles from any land. Here only a continental shelf boundary is required. In this segment the relevant coasts of Norway (in part Svalbard, in part the mainland) and the USSR (Novaya Zemlya) are essentially opposite one another. In the final segment, in the northern Barents, there is need once again for both a continental shelf and economic zone boundary. Here the relevant coasts – Novaya Zemlya and Franz Josef Land on the Soviet side, Svalbard on the Norwegian side – are again opposite. Each of the three segments must be considered separately, both because the applicable law differs and because different factors relevant to delimitation apply to each.

The southern segment

We will begin with the southern segment, which in practice is the area where the need for a boundary is greatest. Before 1977 negotiations in this area were concerned solely with a continental shelf boundary. For this the applicable law was the Continental Shelf Convention (since both states are parties to it). Article 6 of the Convention, it will be recalled, provides that in the absence of agreement, the boundary line is to be the median line unless another line is justified by special circumstances. Since 1977 the negotiations have become concerned with seeking to establish a single boundary for both the economic zone and continental shelf.[34] In this situation is article 6 still the applicable law? In the *Gulf of Maine* case the International Court of Justice held that where a single boundary between overlapping continental shelves and economic zones is to be drawn, article 6 should not be regarded as being applicable even though (as in the *Gulf of Maine* case) all/both the states concerned are parties to the Continental Shelf Convention, because article 6 applies only to one of the zones to be delimited (the continental shelf) and not to the superjacent waters.[35] In spite of this decision in the *Gulf of Maine* case, it appears that Norway and the USSR still regard article 6 as being the governing law in this southern segment.[36] In the discussion that follows article 6 will be treated as the applicable law, although because there is doubt on the matter, reference will also be made to the situation, were customary international law to be the governing law. In any case, it is not altogether easy always to maintain a rigid distinction between article 6 and customary law, especially in the light of the comments of the tribunal in the *Anglo-French Continental Shelf* arbitration that the rules of customary law are a relevant means both for interpreting and completing the provisions of article 6.

Considering first the situation if article 6 were the applicable law: Norway, as we have seen, takes the view that in the light of article 6 the boundary should be the line equidistant from the coasts of Norway and the USSR, and that there are no special circumstances within the meaning of article 6 that call for a modification of the equidistant line – though Norway has said that it is prepared to simplify a strict equidistant line so as to avoid it having too many turning points and segments. Before evaluating the strength of Norway's contention, one preliminary point needs to be dealt with. When maritime boundaries are drawn as equidistant lines, equidistance is usually calculated with reference to the baselines of the states concerned, not their coasts as such. When negotiations over a Barents Sea boundary began and Norway put forward its contention that the equidistance line should be the boundary, the USSR's baseline along its Barents Sea coasts was the low-water mark. In 1985, however, the USSR drew straight baselines along its Arctic coasts, including the Barents Sea.[37] The effect of measuring an equidistant line from these new baselines would be to push the line a little further west in one or two places, though probably not to any marked degree.

Turning now to an evaluation of the Norwegian claim that the equidistant line should be the boundary, the main question to consider is obviously whether there are any special circumstances within the meaning of article 6. The USSR has argued that there are. We will therefore leave this point until we come to evaluate the USSR's arguments below.

If customary international law were the applicable law, then delimitation would have to be in accordance with equitable principles and taking account of all the relevant circumstances so as to reach an equitable result. The Norwegian view would presumably be that the equidistant line is an equitable result. There are, however, two possible arguments against the equidistant line being considered to be an equitable result. The first of these arguments concerns proportionality. As we have seen, international courts and tribunals have generally taken the view that there ought to be a reasonable degree of proportionality (or at least not a significant disproportion) between the lengths of the coastlines of the states concerned, measured in their general direction, and the area resulting to each state from the delimitation.[38] Proportionality is used by international courts, not as a method of delimitation, but as a way of the checking the equitableness of the result of using other methods (in this case, equidistance). According to Tresselt,[39] use of the equidistance line as the boundary in the Barents *as a whole* would divide the continental shelf in the proportion of 60 per cent to the USSR and 40 per cent to Norway. The ratio of each state's share of the lines bounding the Barents as defined in Chapter 1 (so including not only the coastline but lines drawn from the mainland and connecting the various groups of islands), the authors have calculated, is roughly in the proportion 66:33 in the USSR's favour. This calculation was based on drawing, on the Soviet side, a line across the mouth of the White Sea (which is in any case under Soviet law a straight baseline), a line from the eastern end of that line to Kolguyev Island, and thence to the nearest point on Novaya Zemlya. If, however, one included in the Soviet calculation all the mainland coast east of the White Sea to the point where it is opposite the southern end of Vaygach Island (which is between the mainland and Novaya Zemlya), the coastline of Vaygach Island and the southern coastline of Novaya Zemlya excluded from the previous calculation, the ratio would obviously be much more in the USSR's favour – something like 70:30. If one ignored lines joining islands and simply measured the coastline as such (both of islands and mainland), the figure would be even more in the USSR's favour – something like 75:25. This last figure might be regarded as sufficiently disproportionate to the 60:40 ratio of the division of continental shelf to suggest that use of the equidistant line leads to an unequitable result.

However, the figures above relate to the Barents Sea as a whole, while we are concerned here only with the southern sector. We have no figures for applying the proportionality test in this area, and it is difficult to see how such figures could be calculated. Apart from the difficulties alluded to above of measuring the coastlines of the parties along their general direction, there is the difficulty of knowing exactly what areas of continental shelf and economic zone should be included for the purposes of calculating proportionality: for example, how far north and west can one go on the Norwegian side before one is out of this southern sector? On the Soviet side, are the areas of shelf and zone to the east of Kolguyev Island still in the delimitation area? These problems simply underline the point made by most writers on maritime boundary delimitation, namely that in spite of the repeated statements by international courts and tribunals that proportionality is a criterion for assessing the equitableness of any solution, it is a criterion which in most situations it is extremely difficult to apply. In the present case the most that can perhaps be said is that proportionality may raise some slight doubt about the equitableness of a boundary line based on equidistance, and there is a need for cartographers to attempt to come up with rather more accurate calculations than those used above, based on alternative hypotheses as to what are the relevant areas and relevant coastlines.[40]

So much for proportionality: we must now consider the other argument against the equidistance line being considered to be an equitable result. This is the principle of non-encroachment or cut-off. As pointed out above, international courts and tribunals will not regard a boundary line as equitable if it encroaches on, or cuts off, areas that more naturally belong to one of the states concerned rather than the other.[41] In the present case, the Norwegian coast projects a little further seawards than the Soviet coast along what is basically a straight coast between the White Sea and the North Cape. The effect of this is to push the equidistance line a little further to the east, possibly to such an extent that it might be regarded as encroaching on maritime areas that more naturally belong to the USSR, although it is difficult to be conclusive, since in borderline cases like this one, the matter must inevitably be largely one of subjective judgment, be it by a court, the parties or a neutral observer.

To sum up this examination so far of the Norwegian position: some doubts may be raised about the equitableness of using an equidistance line as a boundary line because it may result in a disproportionate division of the continental shelf and economic zone in the southern Barents and because it may encroach on areas which more properly should appertain to the Soviet Union.

Turning back now from customary law to article 6 of the Continental Shelf Convention, what we must now consider is whether there are any special circumstances within the meaning of article 6 which would justify a departure from the median line, as the USSR has argued in this case. These Soviet contentions will now be examined.

The Soviet Union has invoked a whole range of special circumstances to justify a departure from the median line. These include: the configuration and lengths of coastline; geological conditions; the greater population of the Kola Peninsula as compared with northern Norway; ice conditions; the USSR's economic interests; special environmental risks from the Gulf Stream to which Soviet interests are uniquely exposed; and its security interests (which were explained in Chapter 1).[42] The relevance of each of these alleged special circumstances will be examined in turn. As regards the configuration of the coastline, this probably refers to the non-encroachment principle, which has already been discussed. Although the principle has largely featured in the customary rules, it (or more accurately, a coastal configuration which produces encroachment) may equally be considered as a special circumstance within the meaning of article 6. The reference to lengths of coastlines presumably relates to the question of proportionality, which again we have already discussed. As regards geological conditions, whilst at one time international courts and tribunals did regard them as relevant to continental shelf delimitation, the International Court in the *Libya/Malta* case held that because title to the continental shelf was now based on a distance criterion as far as continental shelves within 200 miles of land were concerned (i.e. all states automatically have a continental shelf of 200 miles in breadth as a minimum), geological and geomorphological circumstances were irrelevant in the case (as here) of delimitation of continental shelves within 200 miles of land.[43] Even more is this the case where the delimitation involves (as here) not just a continental shelf boundary, but also an economic zone boundary.[44]

The next Soviet argument, that based on the difference in the size of coastal populations, has not yet been faced by international courts and tribunals. However, given the fact that the trend of their decisions is to exclude as irrelevant to delimitation all factors except the geographical, particularly where, as here, the boundary line relates to both the continental shelf and economic zone, and the fact that arguments based on the differing sizes of land mass of the states concerned have been rejected,[45] it seems likely that a Soviet argument based on differing sizes of population would be regarded as irrelevant. For the same reason, the Soviet Union's ecological and economic arguments would also be regarded as irrelevant.[46] Security considerations are, however, a somewhat different matter. In both the *Guinea/Guinea Bissau* arbitration[47] and the *Continental Shelf (Libya/Malta)* case[48] security factors were not dismissed out of hand as irrelevant, though in neither case was it made very clear in what circumstances they might be relevant: what the court in each case seems to have had in mind is that a state should have control of the maritime territory immediately off its coast. In that sense, security factors would seem to reinforce the non-encroachment principle (see p. 73 above).

Thus, most of the special circumstances invoked by the USSR are irrelevant. The only relevant circumstances are the configuration of the coastline (i.e. the further projection of the Norwegian mainland coastline seawards) and the USSR's security interests, which both refer to the non-encroachment principle, and the differing lengths of the Norwegian and Soviet coastlines, which raise the question of proportionality. As pointed out above, both these matters, particularly the non-encroachment principle, may call for some readjustment to the equidistance line. The USSR has argued that this readjustment should take the form of a sector line:[49] as pointed out above, this is the line of longitude running from the terminus of the existing Norwegian–Soviet frontier to the North Pole, with a slight deviation in the Svalbard area to pass outside the 'box' described in the 1920 Svalbard Treaty (see Map 3.2). Before evaluating the USSR's contention that special circumstances should modify the equidistance line to the sector line, something more must be said about this sector line. This line first made its appearance in Soviet policy in 1926. In that year the USSR issued a decree[50] under which:

> All lands and islands situated in the Arctic to the north, between the coastline of the U.S.S.R. and the North Pole, both already discovered and those which may be discovered in the future, which at the time of the publication of the present decree are not recognized by the government of the U.S.S.R. as the territory of any foreign state, are [hereby] declared territory of the union, [namely in the area] between the meridian 32°4′35″ longitude, East of Greenwich ... and the meridian 168°49′30″ longitude, West of Greenwich.

The westerly meridian referred to in this decree, 32°4′35″, passed through what was in 1926 the terminus of the USSR's frontier on the Barents Sea with Finland. Since then, the USSR's frontier has moved several miles further west and Finland has lost its frontage on the Barents Sea. In spite of this, the sector line referred to by the USSR in its maritime boundary negotiations with Norway appears to be this original 1926 line. Thus, the reference above to its running from the terminus of the Norwegian–Soviet frontier is not quite accurate. On the other hand, the reference to the deviation in the Svalbard area is: this is a modification which is not found in the original 1926 decree but has been introduced by the USSR in the course of the maritime boundary negotiations. It will also be noted that the decree refers quite explicitly only to *land territory*, and says nothing about maritime territory or maritime frontiers.[51] We return now to the contention of the USSR that modification of the equidistance line because of special circumstances should take the form of the sector line, a contention that would seem to have little merit. It would seem highly artificial to consider that special circumstances could cause an equidistance line to be modified to such a precise and, from the Soviet point of view, well-established line.

It seems, however, that as an alternative the USSR has argued that the sector line is itself a special circumstance: the USSR refers to its use in Soviet administrative practice, in a way that gives it a special psychological and political signficance.[52] This argument again has little merit. A sector line is not of the character of circumstances which international courts have hitherto recognised as relevant. It is true that international courts have attached some weight to lines which have been used provisionally as boundary lines, but such provisional boundary lines must have been accepted or acquiesced in by both/all parties to the dispute.[53] That is clearly not the case here, where Norway has consistently objected to the use of the sector line as the boundary.

Finally, turning from article 6 to customary international law, it is possible that the USSR considers that the use of the sector line is justified either because a sector line is the appropriate way of drawing a maritime boundary in polar areas or because a sector line is an equitable result. If the former is the Soviet contention, the USSR would have to prove that the use of sector lines as a method for drawing maritime boundary lines in polar areas is a rule of customary international law and that such a rule is binding on Norway. (Because it is seeking to rely on an alleged rule of customary international law, the burden of proof, as is normal in international law, is on the USSR.) There does not appear to be such a rule as that alleged. In reaching this conclusion the authors have relied heavily on the recent, exhaustive study of the question by Pharand.[54] Briefly, the evidence for this conclusion is that there is both insufficient practice and inadequate *opinio iuris* (state practice and *opinio iuris* being the two constituent elements of customary international law). It is true that sector lines have been used as boundaries in the Antarctic, but this use is limited to *land territory*, and seems to be motivated by reasons of convenience, not because the states concerned felt themselves to be under any legal obligation to use such lines. Nor have all these Antarctic sector lines been recognised by other states. It is also true that two 19th-century treaties – the Great Britain–Russia Convention of 1825[55] and the USA–Russia Convention of 1867[56] – utilise sector lines as boundary lines in the Arctic (although the former, establishing the boundary between Canada and Alaska, is probably limited to a *land* boundary), but again the reason for their use appears to be convenience, rather than any form of legal compulsion resulting from the existence of a rule of customary international law requiring the use of sector lines. Apart from these treaties, sector lines have been used by the USSR in its municipal legislation (notably in the 1926 decree already referred to) and, more ambiguously, by Canada. On the other hand, Norway and the USA have consistently, and Denmark less certainly, objected to the use of sector lines in the Arctic. Thus, even if there were a rule of customary law requiring the use of sector lines as maritime boundaries in the Arctic, this persistent objection by Norway would prevent its being applicable to Norway.

It is also noteworthy that the USSR has refrained from formally endorsing the sector principle for maritime delimitation generally;[57] and that in the provisions of its continental shelf and economic zone legislation dealing with boundaries with neighbouring states (quoted earlier), there is no reference to the use of sector lines. Scrivener suggests that more recent Soviet writers, in contrast to their predecessors, place much less emphasis on the sector principle.[58]

If we turn to the alternative possible justification in customary law of the sector line as the boundary line, namely that the sector line represents an equitable result applying the normal principles of customary law delimitation, this too does not seem very persuasive. An international tribunal would probably not regard the sector line as an equitable boundary in the southern part of the Barents, because such a line would encroach on maritime areas which more properly belong to Norway.

Having now examined the various contentions put forward by Norway and the USSR as to where their maritime boundary should lie in the southern part of the Barents Sea, we have arrived at the conclusion that, even if article 6 of the 1958 Continental Shelf Convention is the applicable law (about which there is doubt), there may be special circumstances, namely the configuration of the parties' coastlines and possibly the USSR's security interests, which would justify some departure from an equidistance line, as such a line risks encroaching on areas that more appropriately belong to the USSR and may cause some disproportion between the length of the parties' respective coastlines and the area of maritime zones resulting to each. In each case, it was suggested above, there is no objective method for determining encroachment or lack of proportionality: it is to a large extent a matter of subjective judgment, which is why no dogmatic conclusion is offered here. If these special circumstances are thought to exist, they would not seem to be of such an extent as to require any very extensive modification of the equidistance line westwards. Certainly, under no circumstances do they require the equidistance line to be modified to the sector line proclaimed by the USSR, nor is there any reason why this latter should autonomously constitute the boundary.

If customary international law were the applicable law then the boundary would be agreed by the parties, or decided by a court or tribunal, by applying equitable principles and taking into account the relevant circumstances so as to produce an equitable solution. If the matter were settled by agreement, then no doubt the parties would regard the result as equitable. If, on the other hand, the matter were decided by a court or tribunal, one can only speculate rather broadly what the result might be. One possible approach of a court or tribunal might be to begin by observing that Norway and the USSR were adjacent states along what is essentially a straight coastline and that the Soviet coastline in the Southern Barents was considerably longer than the Norwegian coastline. What it might then suggest as an equitable boundary might be a line drawn perpendicular to the general direction of the coast. This was commended by the International Court in the *Gulf of Maine* case as a suitable method of delimitation on coasts which are more or less straight.[59] Furthermore, a perpendicular line would respect the principles of non-encroachment and proportionality.[60]

Having now looked at the first segment of the maritime boundary in the Barents, which is in practice the most important, because it is in this part of the disputed boundary area in the Barents that there is most fishing, the greatest likelihood of finding oil and gas and the USSR's security interests are greatest, we must turn to examine the second, middle, segment.

The middle segment

In this segment, it will be recalled, only a continental shelf boundary is required. Thus there is no doubt that the applicable law is article 6 of the Continental Shelf Convention. In other words, in the absence of an agreement between the parties on a different solution, the boundary will be the median line unless another line is justified by special circumstances. Of the various alleged special circumstances raised by the USSR in connection with the boundary in the southern segment and which were found to be relevant in that area, the configuration of the coastline (which was taken to be a reference to the non-encroachment principle) is not relevant to this middle segment of the boundary. This is because the non-encroachment principle has no application where, as here, the respective coasts are opposite one another rather than adjacent. A second relevant circumstance in the southern segment was the respective lengths of the coastlines of the parties (which was taken to be a reference to the proportionality principle). This principle could be relevant to the middle segment of the boundary, but the test of proportionality is at least as hard to apply in this segment as in the southern segment. It would be difficult to calculate the lengths of the parties' relevant coastlines, particularly in the case of Norway, where the relevant coastlines would include part of Svalbard and possibly also part of the Norwegian mainland. Equally, it would be difficult to calculate the area of overlapping continental shelves. Whilst at first sight it might seem that this area was simply the triangular-shaped area beyond the 200-mile limit (see Map 3.2), it is clear on reflection that the Norwegian and Soviet continental shelves also overlap to both the east and west of the triangular-shaped area.

In this middle segment of the boundary the USSR has argued that there is an additional special circumstance. According to the USSR, the Svalbard archipelago should not be given full effect in determining a boundary because of its special legal status.[61] This argument seems to be without any legal merit. It was argued in the previous chapter that Norway's sovereignty over Svalbard means that Norway can claim all the generally accepted maritime zones in respect of Svalbard: there is nothing in the rights given to the other parties to the Svalbard Treaty to limit this aspect of Norway's sovereignty over Svalbard. The same argument would apply equally to the use of Svalbard in general as a coastline for calculating a boundary. On the other hand, there is an argument for saying that some individual islands in the Svalbard archipelago should not be used, or should not be given full effect, in calculating the boundary. This is not because of anything to do with the legal status of Svalbard but follows from the general rules of maritime boundary delimitation: as pointed out above, there has been a trend in the practice both of international courts and tribunals and of states in concluding bilateral delimitation agreements to discount or give only partial effect to isolated islands. In the case of Svalbard, several islands – Hope Island, the King Karl's Land group of islands and Bear Island – lie a long way from the main part of the archipelago. These islands should thus not be used as basepoints in calculating an equidistance line. Instead, they should be either ignored completely or given partial effect (i.e. the equidistance line should be measured from a point somewhere between these islands and the main Svalbard archipelago). Finally, it should be stressed, as indeed should be clear, that the sector line is no more a special circumstance or mandatory boundary than it was in the southern segment.

To sum up the position in this middle segment of the Barents Sea maritime boundary: because the boundary here is a continental shelf boundary, unless the parties agree otherwise, the boundary should be the median or equidistance line, modified so as to discount, wholly or partially, Hope Island, King Karl's Land and Bear Islands.

The northern segment

In this part of the Barents, it will be recalled, there is a need, as in the southern Barents, for both a continental shelf and economic zone boundary. Again, there is the same doubt as to what is the applicable law, article 6 of the Continental Shelf Convention or customary international law. If article 6 is the applicable law, then in the absence of an agreement between the parties to the contrary, the boundary will be the equidistance line, unless another line is justified by special circumstances. In this area the relevant coasts are the Svalbard archipelago on the Norwegian side and the Franz Josef Land archipelago and Novaya Zemlya on the Soviet side. Of the various alleged special circumstances raised by the USSR and discussed in relation to the other two segments of the boundary, the only relevant circumstance is the configuration of the coast. There are various outlying islands in both the Svalbard and Franz Josef Land archipelagos that ought to be wholly or partially discounted in calculating the equidistance line.

If the applicable law is customary international law, the boundary is likely to be much the same. This is because in general international courts have taken the view that equidistance is an appropriate method of delimitation (at least as a starting point) in the case of opposite coasts.[62] Again there would need to be some adjustment to a strict equidistance line to discount the outlying islands of the Svalbard and Franz Josef Land archipelagos.

In relation to this northern segment, two further points are worth noting. First, the presence of ice-shelves off parts of Svalbard (especially off the south-east coast of North-east Land) and Franz Josef Land may complicate the drawing of an equidistance line, as there may be differences of view between Norway and the USSR as to whether an equidistance line should be calculated from the land or from the outer edge of the ice-shelf. The question of whether ice-shelves can be used in delimiting maritime zones is unsettled in international law, and the views of writers are divided as to whether their use is or should be permissible.[63] Second, in part of this northern area, the sector line follows very closely the equidistance line (see Map 3.2). Even though this is of no legal relevance (because, as stressed earlier, the sector line has no standing in international law), it should make it easier for the parties to agree on the equidistance line, or some modified version of it, as the boundary.

Having examined the parties' arguments as to where their maritime boundary in the Barents Sea should be drawn in the light of the applicable international law, we will now turn to consider what the prospects are for a solution to this dispute.

TOWARDS A SOLUTION?

The solution to the Barents Sea boundary dispute has both a procedural and a substantive aspect; that is, the question of what procedure will be used to arrive at a solution, and the question of what the actual substance of any solution will be. As far as the former is concerned, until very recently one would have been certain that the procedure by which the dispute would be resolved, if it were to be resolved, would be by continued bilateral negotiations between the parties. The alternative to negotiations is some form of third party settlement – conciliation, arbitration or use of the International Court of Justice. The use of such forms of third party settlement is purely optional in international law – no state can be forced to go to arbitration or the International Court of

Justice against its will. Traditionally the USSR has always opposed the third party settlement of disputes. However, this aspect of Soviet policy has also been affected by Mr Gorbachev's policies of *glasnost* and *perestroika*. The USSR has recently proposed that greater use should be made of the International Court of Justice for settling disputes and has withdrawn its reservations to the clauses in various human rights treaties to which it is a party, conferring jurisdiction on the Court to deal with disputes arising under those treaties.

Nevertheless, in spite of this radical change in Soviet attitudes to the third party settlement of disputes, it is still likely that a solution to the Barents Sea maritime boundary dispute will come through direct negotiations and not some form of third party settlement. Like most states, Norway and the USSR are likely, in the light of past practice and the vagueness of the relevant international law rules, to regard the outcome of a submission of their dispute to arbitration or the International Court as being too unpredictable. It is true that Norway's dispute with Denmark over the maritime boundary between Jan Mayen and Greenland is currently before the Court, but this is contrary to Norway's wishes: Norway would have preferred the dispute to be resolved by negotiation. The case was referred to the Court by Denmark, relying on its and Norway's long-standing declarations under article 36(2) of the Court's Statute (the optional clause) to found the Court's jurisdiction. In the case of the Barents Sea boundary dispute, the USSR is likely especially not to wish to refer the dispute to third party settlement, because of the implications of the dispute and any solution for its security interests.

Even if direct negotiations seem the most likely procedure by which the dispute will be resolved, it is necessary to add a few words about the possible significance of the 1982 UN Convention on the Law of the Sea, should the Convention enter into force and Norway and the USSR be parties to it before the dispute is resolved. Part XV of the Convention provides in general for the compulsory third party settlement of disputes. However, article 298(1)(a) provides that a state may, when signing or ratifying the Convention, declare that it does not accept the third party settlement of dispute procedures under the Convention (which are arbitration or the use of the International Court of Justice or the International Tribunal for the Law of the Sea) in the case of maritime boundary disputes. Nevertheless, a state having made such a declaration must accept compulsory conciliation for 'disputes concerning the interpretation of articles 15, 74 and 83 relating to sea boundary delimitations . . . when such a dispute arises subsequent to the entry into force of this Convention', if any party to the dispute so requests. Although this provision is not without a certain ambiguity, it is clear from the *travaux préparatoires* of the Convention that the intention is to exclude maritime boundary disputes arising *before* the entry into force of the Convention from the obligations of compulsory conciliation.[64] When signing the Convention, the USSR made a declaration under article 298 (1)(a). If it maintains this declaration when it ratifies the Convention, or if Norway makcs a similar declaration when it ratifies the Convention, then, because the Barents Sea dispute arose before the entry into force of the Convention, there will be no obligation to refer the dispute either to the forms of third party settlement established by the Convention or to compulsory conciliation. If, as seems unlikely, both the USSR and Norway were to refrain from making a declaration under article 298(1)(a), then there would be an obligation on them to refer their dispute, if it still could not be settled by negotiations, to arbitration, the International Court or the Law of the Sea Tribunal. Thus, the most likely result is that the entry into force of the Law of the Sea Convention would not affect the procedure by which the Barents Sea maritime boundary will be resolved, which will continue to be direct negotiations between the parties.

The speed at which negotiations over a maritime boundary in the Barents are continued and the likelihood of their reaching a successful conclusion depend to a considerable extent on how strongly the parties feel the need to explore new areas for sea-bed hydrocarbons. As long as the 'Grey Zone Agreement' remains in force, fisheries interests will not press either party to seek an early solution to the boundary problem. As explained in Chapter 5, there are a number of reasons why both Norway and the USSR in the fairly near future are likely to want to step up their exploratory activities for oil and gas in the Barents Sea (although at the same time there are a number of factors that are likely to slow down such a development). While the respective desires of the parties to explore for hydrocarbons in the Barents Sea is likely to be a major factor in determining the speed and course of negotiations, it is not the only factor. For the USSR security considerations are also very important. Its security interests would, in the absence of any pressing need for resources, lead the USSR to delay negotiations. So long as there is no agreed boundary, the Norwegians will not drill in the disputed area or in all probability so very close to it. Wherever a boundary is located, it will almost certainly lead to Norwegian offshore activity pushing eastwards, something which is inimical to Soviet security interests: the further east Norwegian offshore activity takes place, the greater the possibility of Norwegian drilling rigs obstructing or monitoring Soviet naval vessels on their way into or out of the ports of the Kola

Peninsula.[65] A final factor in determining the speed of negotiations is the fact that the USSR may perceive Norway to be keener on a boundary agreement than it is: it can therefore afford to wait, hoping that this will lead to concessions by Norway.[66] However, given the rapid pace of change currently taking place in the USSR, it is unwise to be too dogmatic with any predictions about how long it will be before negotiations achieve a positive result.

It is, of course, equally unwise to be too dogmatic about predicting what this result might be. While some suggestions have been made in the preceding section as to the possible course of the boundary indicated by international law, it should be stressed again that it is open to the parties, whether the applicable law is article 6 of the Continental Shelf Convention or customary international law, to agree on any boundary line they wish. It may also be that the parties would be wise to broaden their negotiations beyond simply searching for agreement on a boundary line. A number of writers have suggested that jurisdictional disputes and disputes over maritime boundaries are likely to be more easily settled if one abandons discussing such disputes purely in terms of reaching a definite solution on jurisdiction or an agreed boundary, and instead looks at the issues underlying the dispute and attempts to address these: in other words, one to some extent redefines the problem.[67] Looked at in this way, the issues in the Barents Sea maritime boundary dispute are, from the Soviet point of view, its security concerns and a desire for a reasonable share of the natural resources of the Barents (oil, gas and fish), and, from the Norwegian point of view, a similar desire for a reasonable share of natural resources and a wish not to be seen to have given in to the USSR or entered into any kind of sharing or condominium arrangement with it. Norway's security concerns are much less of a factor because its security interests (which were explained in Chapter 1) are not likely to be directly affected by the location of the boundary (at least on the assumption that it will not be further west than the sector line). In the light of this analysis, one might therefore suggest that the parties should search for a boundary settlement having the following five elements, the first two of which are (1) agreement on a boundary line, probably lying somewhere between the median line and the sector line; and (2) a way of addressing the USSR's security interests.

Offshore oil and gas activities in the Norwegian sector of the Barents Sea raise a number of possible problems for the USSR from a security point of view.[68] First, oil and gas installations pose an obstacle to navigation by Soviet naval vessels, both surface vessels and submarines,[69] particularly since the effect of ice and sea-bed conditions is to channel Soviet naval vessels out of the Barents between Bear Island and mainland Norway – the area of the Norwegian sector most likely to see oil and gas activities. Second, installations have some espionage potential, although this is limited because the noise they produce makes them not very suitable listening-posts for submarines. Third, the increase in the background noise underwater resulting from both installations and other activities, such as seismic testing, makes it more difficult for the USSR to detect non-Soviet submarines through underwater listening devices – though equally, of course, it makes it more difficult for Soviet submarines to be detected.

To deal with the first two of these problems, a boundary settlement could include designating an area on the Norwegian side of the boundary line where the USSR would be notified and/or consulted over the emplacement of oil and gas installations, in order to avoid major interferences with Soviet naval shipping, and offering the USSR a guarantee that Norwegian installations would not be used for espionage (even though this might raise difficult questions of verification and inspection). To some extent such arrangements would be a fairly natural extension of and building on the existing agreement between Norway and the USSR on the prevention of incidents at sea, signed in December 1989. That the USSR might be receptive to such ideas is suggested by Mr Gorbachev's Murmansk speech, which called for restrictions on various forms of naval activity in northern and Arctic waters as a confidence-building measure. As regards the third problem, increased noise, there seems no effective way of dealing with this without prohibiting or drastically limiting offshore oil and gas activities in the Barents Sea, something neither Norway nor the USSR would be likely to agree to. In any case, as pointed out above, increased noise has advantages for the USSR as well as disadvantages.

The other three elements would be: (3) a continuation and possible modification of existing arrangements relating to the joint management of fish stocks in the Barents (which are discussed in the next chapter); (4) arrangements for the unitisation of any oil or gas fields straddling the boundary line, which Norway has already proposed[70] and numerous examples of which exist elsewhere – such arrangements are, or course, quite different from the zone of joint development which the USSR has proposed and Norway has rejected; and (5) arrangements providing for co-operation in the event of any pollution resulting from offshore oil and gas activities. Such arrangements are to some degree in fact already being discussed by the Joint Norwegian–Soviet Commission on Environmental Co-operation, established by the Agreement between Norway and the USSR on Environmental Co-operation of 1988. This matter is considered in further detail in Chapter 5.

The comments made by the Soviet Prime Minister, Mr Ryzkhov, on his visit to Oslo in January 1988, referred to earlier, suggest that the USSR might be responsive to such an approach to resolving the dispute.

In addition, or as an alternative, it is possible to envisage a further broadening of the issues to involve the unresolved questions concerning Svalbard (discussed in Chapter 2) in order to arrive at a giant package deal involving both Svalbard and the Barents Sea boundary. Such a possibility, which has been advocated by some writers,[71] will be discussed in Chapter 6.

We have now explored most of the issues concerned with the delimitation of maritime boundaries in the Barents Sea, but there is one final question which must be examined, and that is whether there is any obligation on Norway and the USSR in international law not to explore or exploit the sea-bed of the disputed area, pending agreement on a boundary.

IS THERE ANY PROHIBITION IN INTERNATIONAL LAW ON EXPLORING AND EXPLOITING THE DISPUTED AREA?

Given the length of time negotiations over a maritime boundary in the Barents Sea have lasted and may continue to last, given the discovery of oil and gas in the eastern Barents and the expansion of Norwegian oil and gas activities northwards, it is obviously an important practical question as to whether international law prohibits Norway and the USSR from unilaterally engaging in any activities relating to the exploration and exploitation of sea-bed resources until agreement has been reached on a boundary. While a clear-cut answer cannot be given to this question, there is quite a lot of evidence to suggest that there is a rule of customary international law prohibiting states from unilaterally engaging in offshore hydrocarbon activities in disputed areas of the continental shelf.

In the first place, there are a number of broad principles of international law providing support for such a rule. Under article 6 of the Continental Shelf Convention a continental shelf boundary 'shall be determined by agreement' between the states concerned. This implies that such states are under an obligation to enter into negotiations with a view to reaching agreement. This is also the position in customary international law, as the International Court of Justice stressed in the *North Sea* cases.[72] It is a rule of international law, as has been stressed on many occasions by international courts and tribunals,[73] that negotiations must be conducted in good faith. It can be argued that to drill in a disputed area of continental shelf subject to delimitation negotiations is a breach of good faith because it is an action which may be regarded as trying to prejudice the outcome of the negotiations. Since only a coastal state can explore and exploit its continental shelf, such action would be tantamount to asserting that the area of disputed continental shelf where drilling had taken place belonged to the state undertaking the drilling. If there is such a principle, there is some difficulty in ascertaining its temporal application. While there is an obligation to negotiate in good faith, there is no obligation to reach an agreement. At some stage, therefore, if agreement cannot be reached, the obligation to negotiate must lapse. This would imply that the obligation not to drill in the disputed area would also lapse.

A second, and to some extent related, argument is as follows. A coastal state has the exclusive right to explore and exploit the resources of its continental shelf. Therefore, if state A explores and exploits resources of a disputed area of continental shelf which later turns out to belong to state B, it will have committed an international wrong against state B. (State B's rights over its continental shelf, it should be noted, exist *ab initio* and not simply from the date of the delimitation when its title to the disputed continental shelf becomes unequivocally established.) It could be argued that state A is under an obligation (deriving again from good faith) not to risk committing such a wrong and thus must not drill in the disputed area. While some writers have asserted that this must be so,[74] the matter does not seem entirely free from doubt, particularly in the light of an *obiter dictum* in the *Aegean Sea* case. In this case the International Court of Justice held that seismic surveys by Turkey in an area of continental shelf claimed by both Turkey and Greece did not constitute such an irreparable prejudice to Greece's rights that interim measures of protection were justified. The Court added, however, that seismic exploration of a continental shelf without the consent of the coastal state 'might, no doubt, raise a question of infringement of the latter's exclusive right of exploration'. If so, and if Turkey's exploration took place in an area that was subsequently adjudged to be Greek continental shelf, there would be a case for reparation by Turkey to Greece.[75] Thus, the Court contemplates that a wrong could be committed for which reparation would subsequently be made: it does not suggest that there is any obligation to avoid the risk of committing a wrong in the first place, as was argued above might be the case, although it may be that the Court's comments should be regarded as being limited to the situation where the exploration is by means of seismic tests. It might have taken a different view had drilling been involved. While the Court's statement is only an *obiter dictum* and its exact scope is unclear, it does cast doubt on the argument that a state must not drill in a disputed area of continental shelf in order to avoid the risk of committing a wrong against another state.

To the above two broad principles suggesting that a state is prohibited from unilaterally exploring and exploiting a disputed continental shelf area may be added another. Lagoni persuasively argues, relying on a dictum from the *Electricity Company of Sofia and Bulgaria* case[76] and certain provisions of the General Act for the Pacific Settlement of Disputes of 1928,[77] that there is a general rule of customary international law that parties to a dispute must not take any steps that would aggravate or extend the dispute.[78]

Apart from these general principles, there are a number of examples of state practice which provide further evidence for there being a rule of customary international law that a state may not unilaterally engage in offshore oil and gas activities in disputed areas of continental shelf. First, Tunisia and Libya halted offshore oil and gas operations in the area of disputed continental shelf while their case was being dealt with by the International Court of Justice.[79] Second, the USA in 1976 and 1978 withdrew from consideration areas proposed for offshore hydrocarbon licensing because they were situated on part of the continental shelf on Georges Bank also claimed by Canada.[80] Third, Greece and Turkey in 1976 signed an Agreement on the Procedure to be followed for the Delimitation of the Continental Shelf,[81] paragraph 6 of which provides that 'both parties undertake to abstain from any initiative or act relating to the continental shelf of the Aegean Sea which might prejudice' their negotiations over a continental shelf boundary. Fourth, following exploratory drilling for oil by Denmark in 1983 in an area of continental shelf in the Kattegat also claimed by Sweden, the Swedish government sent the Danish government a note of protest in which it claimed that the Danish action was contrary to international law. Subsequently, Denmark and Sweden issued a communiqué in which they agreed that 'activity in the disputed area should be based on the principle of having regard to the other party's interests, including the principle of consultation between the parties'. It was also stated that neither party would permit exploratory drilling during the period of negotiations over a boundary.[82] Lastly and quite significantly in the present context, in 1982 the Norwegian Minister for Oil and Energy stated in reply to a Parliamentary question that 'international law requires mutual restraint. The outcome of the boundary negotiations [with the USSR] must not be anticipated through unilateral action.'[83]

From these quite extensive instances of state practice, coupled with the general principles discussed earlier, it seems reasonable to conclude that there is probably a rule of customary international law requiring states not to engage unilaterally in offshore oil and gas activities in disputed continental shelf areas. However, it is not altogether clear exactly what activities are covered by this probable rule. Certainly, it would seem that exploratory drilling is covered, as of course is production. But what about seismic tests? Here there must be more doubt. The general principles discussed earlier appear to apply to all the continental shelf rights of a coastal state, which include its exclusive sovereign rights of exploring and exploiting the resources, as well as its right to regulate research on its continental shelf. The latter at least covers seismic tests, even if exploring possibly does not (this point was left open in the *Aegean Sea* case). This would therefore mean that seismic tests were included in the prohibition on unilateral acts. On the other hand, the *Aegean Sea* case casts doubt on this; and the examples of state practice shed no light on the matter. The question of whether seismic tests are included in the probable rule therefore remains uncertain. However, since seismic testing involves substantially less interference with the sea-bed than drilling, there are good reasons why it might be considered differently and excluded from the scope of the rule.

The basic rule itself will be strengthened if and when the 1982 UN Convention on the Law of the Sea enters into force. Articles 74(3) and 83(3) provided that pending the reaching of agreement on economic zone and continental shelf boundaries the states concerned 'shall make every effort ... not to jeopardise or hamper the reaching of the final agreement'. Again, it is not clear exactly what activities are caught by this provision. Lagoni persuasively argues that it covers 'any activity which represents an irreparable prejudice to the final delimitation agreement', including exploratory drilling but not exploratory activities of a transitory character such as seismic tests, unless in exceptional circumstances they would aggravate the dispute.[84]

Having seen that there probably is a rule of customary international law (which is closely mirrored in the Law of the Sea Convention) which prohibits unilateral oil and gas activities on disputed continental shelves (even if the exact activities encompassed by this prohibition remain unclear), it remains to examine how far Norway and the USSR have complied with such a rule in the disputed sea-bed area in the Barents Sea. The short answer is that on the whole the rule has been complied with. On the Soviet side, the one breach of the rule was in May 1983 when the drilling ship, *Valentin Shashin*, drilled about 1.5 miles west of the median line inside the disputed area, but it later transpired that this transgression had been accidental: in any case the ship was within the

2–3 mile margin of error corridor recognised by Norway.[85] Both the USSR and Norway have engaged in seismic testing in the disputed area, and each has protested to the other.[86] However, as we have seen, there is doubt as to whether seismic testing is covered by the prohibition on unilateral acts in disputed continental shelf areas.

CONCLUSIONS

As we have seen, Norway and the USSR have been negotiating over a maritime boundary in the Barents Sea for nearly twenty years. Norway has argued that the boundary should be a median line, the USSR that it should be a sector line. Because of the vagueness and generality of the rules of international law relating to maritime boundary delimitation, international law does not prescribe with anything approaching precision where the boundary should lie. Nevertheless, there is no justification in international law for the sector line being the boundary. There is more support for the median line, especially if modified, first, to take account of the further projection of the Norwegian mainland coast seawards; second, to reflect a reasonable degree of proportionality between the lengths of the parties' respective coastlines and the areas of continental shelf belonging to each; and third, to discount, wholly or partially, the effect of various off-lying islands in the Svalbard and Franz Josef Land archipelagos. Ultimately, however, the boundary is a matter for negotiation between the parties (unless, as seems very unlikely, recourse is had to some form of third party settlement). In such negotiations, the parties are, from a power-politics point of view, in an unequal bargaining position, which is accentuated by the lack of precision and of the obligatory nature of the legal rules on which the weaker party (Norway) can rely in support of its case.[87]

In spite of the length of time that the parties have been negotiating over a boundary, there has appeared until recently to have been very little narrowing of the two parties' original positions. Norway has indicated that it is prepared to make concessions on the median line, whereas the USSR appeared to stick uncompromisingly to the sector line, without offering any concession, until 1988, when it proposed a zone of joint development as at least a temporary solution to the boundary problem, a proposal Norway immediately rejected because of its desire to avoid any kind of condominium arrangements in the far north. More recently there have been some further signs that the USSR may be prepared to modify its original position. Nevertheless, there seems little likelihood of any solution to the boundary dispute in the near future, although given the current capacity for unforseen changes in all matters involving the USSR, it would be most unwise to make any firm predictions as to when a boundary solution might be reached and what its substance might be.[88] There seems little doubt, however, that if the location of the boundary did not have considerable security implications for the USSR, a maritime boundary would probably have been agreed by now.[89] Thus, as suggested earlier, agreement on a boundary is more likely to be reached if it is included as part of a settlement that addresses the USSR's security interests.

During the considerable length of time the boundary dispute has lasted, the parties have succeeded in keeping the dispute low-key, therefore not adding greatly to the existing tension in the area. It is true that there have been occasional incidents which could be seen as connected with the dispute, such as the USSR engaging in weapons tests in part of the disputed area in 1976 and 1979 and in 1985 cutting the streamer of the Norwegian ship, *Malene Østervold*, when engaged in seismic tests in an area of continental shelf well to the west of the sector line and therefore outside the disputed area, but such incidents have been few and far between. In particular, with one accidental exception, the parties have observed a probable international law obligation not to drill for oil and gas in the disputed area pending agreement on a boundary. As far as fisheries are concerned, the 'Grey Zone Agreement' has provided a reasonably satisfactory temporary solution to the absence of an agreed boundary. Thus, as maritime boundary disputes go, this has so far been one of the better-managed ones. It is to be hoped that this state of affairs will continue until the dispute is solved.[90]

4 Fisheries management and access to fishery resources

INTRODUCTION

The Barents Sea and the Norwegian Sea are regarded as forming a single ecosystem, and this maritime area is biologically one of the most productive in the world. The average annual catch of the most important fisheries in this area has been 2.2 million tonnes for the last twenty years.[1] Of the total Norwegian catch in 1978–85, 72 per cent was taken in this area, and fishing for the Barents Sea stocks is of vital importance to the northernmost regions of Norway. The USSR catch is about the same as the Norwegian in quantity, but this represents a smaller percentage of the total Soviet catch. (Further information about the fishery resources of the Barents Sea was given in Chapter 1.)

Traditionally, the rich fishing grounds in the Barents Sea have also been exploited by fishermen from other nations. In 1976, before the establishment of 200-mile zones, 20 per cent of the most important fish stock, the North-east Arctic cod, was caught by states other than Norway and the USSR. The most important of these other fishing nations were states from the European Community, especially the United Kingdom, the Faroe Islands and the Eastern European states, Poland and the German Democratic Republic.[2]

As in other parts of the world, a regional fisheries organisation was established, the North-East Atlantic Fisheries Commission (NEAFC) of 1959.[3] The Commission became operative from 1963 and its regulatory powers applied not only to the high seas but also to the territorial seas and the then twelve-mile national fishing zones of the North-east Atlantic (including the Barents Sea). The Commission was authorised to make recommendations by a two-thirds majority vote (art. 8(1)), but each of the member states could reject these recommendations by submitting a protest within ninety days (art. 8(2)). Recommendations were enforced either by coastal states within their maritime zones or by flag states on the high seas.

Originally the Commission's powers were limited to making recommendations concerning the mesh sizes of nets and minimum fish sizes, and closed seasons and closed areas. In 1974 the Commission's powers were considerably enlarged when it was authorised to regulate the total catch. The way in which the Commission did this was by establishing a Total Allowable Catch (TAC). The TAC was then divided into quotas which were allocated among interested member states. In order to improve enforcement on the high seas, the Commission in 1969 adopted a Scheme of Joint Enforcement.[4] This permitted authorised inspectors to board the vessels of any member state on the high seas and inspect their nets and catch in order to verify compliance with the Commission's recommendations. Any violations discovered were reported to the flag state, which alone could take proceedings to punish the master of the offending vessel.

The NEAFC, however, was not able to reach agreement on necessary conservation measures. In particular, TACs set by NEAFC often exceeded those recommended by the International Council for the Exploration of the Sea (ICES), on which the NEAFC relied for scientific advice. For example, in 1975 the ICES recommended a TAC for North-east Arctic cod of 625,000–650,000 tonnes. The actual TAC adopted by the NEAFC was 810,000 tonnes. Furthermore, there was heavy over-fishing: for example, the actual catch of North-east Arctic cod in 1975 was 834,000 tonnes. This development resulted in a depletion of fish stocks, which occurred as much in the Barents Sea as anywhere in the North-east Atlantic. An evaluation of major stocks in the North-east Atlantic by 1975–6 classified two stocks as under-exploited, one as fully exploited, twenty-eight as over-exploited and two as depleted.[5] This development is an example of what has been called 'the tragedy of the commons'. This concept is used to describe the phenomenon whereby an open access regime tends to promote competition for resources, in this case fish, leading to their eventual depletion.[6]

THE ESTABLISHMENT OF 200-MILE ZONES

Due in part to the weaknesses of the NEAFC regime, there was a strong political pressure in Norway to extend its fishing limits beyond twelve miles. On 26 September 1974 the Norwegian government declared four principal policy aims:[7] (1) To establish trawler-free zones beyond the 12-mile limit in 1974; (2) To establish a 50-mile fishing limit off Northern Norway in 1975; (3) To promote the concept of 200-mile Exclusive Economic Zones (EEZs) at the Third United Nations Conference on the Law of the Sea (UNCLOS III); and (4) To establish Norwegian measures on the basis of international law and after contact and negotiations with the states concerned.

The USSR was, due to her build-up of a large distant-water fishing fleet, initially sceptical of accepting coastal state fisheries jurisdiction in 200-mile EEZs.[8] But in 1974 there was a major shift in Soviet policy. The USSR made it known that it was prepared to accept the 200-mile EEZ, but only, however, as part of a compromise at UNCLOS III, and not as unilateral measures by coastal states.[9]

In pursuance of the policy to extend its fisheries jurisdiction, Norway entered into negotiations with affected states to establish trawler-free zones beyond the 12-mile limit. The negotiations resulted in treaties with the USSR, the United Kingdom, the German Democratic Republic, Poland, France and the Federal Republic of Germany.[10] Three trawler-free zones were established outside the Norwegian 12-mile limit off northern Norway on 31 January 1975; two of the zones were located in the Barents Sea.[11] But as a result of its negotiations with the USSR, Norway had to accept a limitation of both the geographical extent and the time period of the prohibition against trawling, and one of the four zones originally proposed was dropped.[12] Furthermore, full enforcement authority to Norway was not explicitly accepted by the other states. The treaties with the Western European states accepted Norwegian boarding and inspection of their fishing vessels and the expulsion from the zones of vessels which had committed infringements there. The treaties with the Eastern European states, however, only stated that each of the states was to implement measures to ensure that the treaty was not violated.

Norway and the USSR also entered into a Treaty on Co-operation in the Fishing Industry in 1975,[13] establishing a joint fisheries commission composed of representatives from both states. The treaties between Norway and the USSR on trawler-free zones and on fisheries co-operation should be seen as a *quid pro quo*. The link is obvious when noticing that the ministers of the two states initialled the treaty on co-operation on 13 December 1974[14] and the trawler-free zones were established on 31 January 1975.[15] The two treaties were a victory for Norway in the sense that trawler-free zones beyond the fishing limit were accepted. To negotiate a treaty with the USSR on the trawler-free zones, instead of their implementation on a unilateral basis, was also in accordance with the Norwegian principle of promoting low tension in the North.[16] But the zones were less extensive than desired by Norway, and coastal state enforcement in these zones was not expressly accepted. Norway, which has always been reluctant to accept any sort of joint Norway–USSR management in the North, also had to accept the establishment of a joint fisheries commission. This commission was, however, only to make recommendations, not to establish binding measures.

Norway never established a fifty-mile fisheries limit, the second of its 1974 policy aims, for three reasons. First, in 1974, in a case between the United Kingdom and Iceland, the International Court of Justice did not accept Iceland's then recently proclaimed fifty-mile limit. Second, the states which would have been affected by it did not support a Norwegian fifty-mile limit. Third, and very importantly, 200-mile EEZs were, through UNCLOS III and state practice, starting to emerge as customary international law.

The USSR remained sceptical of the establishment of 200-mile EEZs by unilateral coastal state action. On 30 April 1976 it protested against the establishment of a 200-mile zone by the USA.[17] But on 19 May 1976 the USSR and Canada agreed on an extension of Canada's fisheries limit to 200 miles. Norway and the USSR entered into a fisheries treaty based on 200-mile zones on 15 October 1976.[18] On 10 December 1976 a 200-mile fishery zone was established by the USSR[19] and extended to the Barents Sea on 24 February 1977, with effect from 1 March 1977.[20] This legislation was, however, of an interim nature and was, following the USSR signature of the 1982 United Nations Law of the Sea (LOS) Convention, repealed by the 1984 Edict on the Economic Zone.[21] The Norwegian 200-mile economic zone was established on 17 December 1976, with effect from 1 January 1977.[22]

The 1976 treaty between Norway and the USSR was crucial in three respects: it recognised coastal state fisheries jurisdiction in the 200-mile zone, it established a framework for co-operation on joint stocks and it gave mutual access to the two states' zones. This is another example of Norway, in pursuit of its policy of maintaining a low level of tension, seeking to solve problems by treaty rather than by unilateral steps.[23] But although the Norwegian 200-mile zone was accepted by the USSR, the latter was still a bit reluctant, as it did not accept an explicit reference in the Treaty to the 200-mile zone being a part of customary international law. Neither did it accept inclusion of an enforcement article similar to the one in the then draft LOS Convention. Thus, the Treaty's article 5 only refers to coastal state enforcement action as being 'in accordance with international law'.

THE MANAGEMENT FRAMEWORK

Article 56(1) of the LOS Convention establishes the coastal state's sovereign rights for the purpose of 'exploring and exploiting, conserving and managing' the living resources in the 200-mile zone. These sovereign rights are considered to be part of customary international law.[24] This means that the two coastal states of the Barents Sea, Norway and the USSR, have exclusive jurisdiction over the living resources in their 200-mile zones.

Some of the fish stocks of the Barents Sea, such as saithe, redfish and Greenland halibut, are essentially found within either the Norwegian or the Soviet zones, and are called exclusive resources. But in the North-east Atlantic 80 per cent of the fish stocks are shared between different coastal states' zones.[25] In the Barents Sea, the most important fish stocks, such as the North-east Arctic cod and haddock and capelin, are shared between Norway and the USSR. The Norwegian spring-spawning herring used to migrate between what is now Norwegian, Icelandic and Soviet 200-mile zones. Figure 4.1 shows the migration area of the Norwegian spring-spawning herring before 1970. During its state of depletion, in the 1970s and early 1980s, it was, however, only to be found within the Norwegian zone. More recently, it has begun to resume its former pattern of migration. The migration areas of the main shared Barents Sea stocks today are shown in Figure 4.2. Shared fish stocks necessitate co-operation over management between the owner states. Such co-operation over the management of shared fish stocks is also an obligation under the LOS Convention (art. 63). This means that Norway and the USSR are required to co-operate over the management of the North-east Arctic cod and haddock and of capelin. Since also Iceland has an interest as an 'owner' state in the management of the spring-spawning herring, all three states should be required to co-operate over the management of this stock.

The remaining pocket of high seas in the Barents Sea (see Map 3.2) is not important from a fisheries point of view because a viable economic fishery in this area is not feasible without access to the adjoining 200-mile zones.[26] The reconstituted NEAFC,[27] of which Norway and the USSR are both members, has mainly a role to play in areas that are still high seas, which means that this organisation is now playing only a very minor role in the Barents Sea.

Accordingly, except for spring-spawning herring, Norway and the USSR have in practice the sole responsibility for fishery management in the Barents Sea. Since the introduction of the two states' 200-mile zones, co-operation over management has been conducted within the framework established by the 1975 and 1976 treaties referred to above.

This co-operation is executed through the joint commission established by the 1975 treaty. The commission usually meets once a year to consult over the management and allocation of fishing rights for

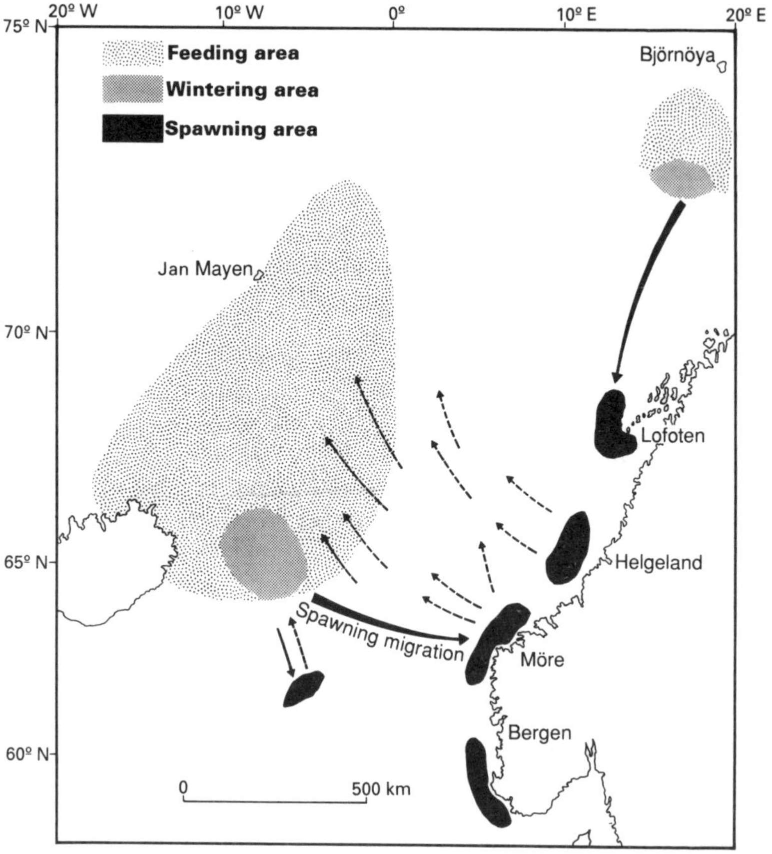

Figure 4.1 Migration area of the Norwegian spring-spawning herring before 1970

Source: J. Hamre, *Fiskebestandene i norske farvann. Naturgrunnlag og beskatning*. Tromsø, Norwegian School of Fisheries, 1986, p. 38

the coming year. These consultations are based on recommendations from the International Council for the Exploration of the Sea (ICES). The 1976 treaty applies to the areas beyond twelve miles, where the two coastal states exercise fisheries jurisdiction (arts 1 and 2). This means that the treaty also applies to the fishery protection zone around Svalbard and to the disputed area in the Barents Sea (the 'Grey Zone'). The commission consults on the management of all stocks within the treaty area, except for whales and salmon. The management of whales is

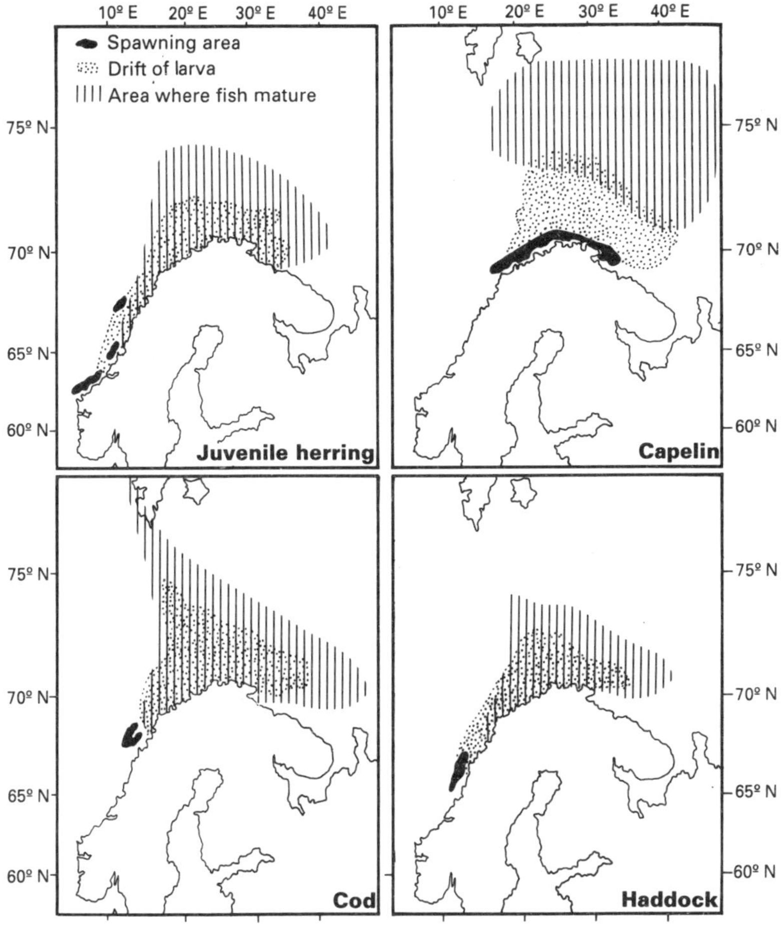

Figure 4.2 Migration area of the most important fish stocks in the Barents Sea

Source: J. Hamre, *Fiskebestandene i norske farvann. Naturgrunnlag og beskatning*. Tromsø, Norwegian School of Fisheries, 1986, p. 38

left to the global International Whaling Commission, while salmon fishing is managed by the regional 1982 Convention for the Conservation of Salmon in the North Atlantic Ocean (see pp. 120–3 below). The catching of seals used to be regulated under an agreement of 1957 between Norway and the USSR,[28] but since 1983 the management of seals has been transferred to the joint fisheries commission. As stated above, the commission only adopts recommendations, not binding decisions.

ALLOCATION OF FISHING RIGHTS BETWEEN NORWAY AND THE USSR

The management regime established by the 1982 LOS Convention may be adequate for exclusive fish stocks, but it is not well suited to shared stocks. The Convention requires each coastal state to establish a total allowable catch (TAC) in its 200-mile zone (LOSC, art. 61(1)). Setting TACs in this way for shared stocks could, however, provide too rigid a fishing pattern. Norway and the USSR agree instead every year upon a TAC for the whole migration area of these stocks. Although being in contradiction to the wording of article 61(1) of the LOS Convention this practice should be regarded as legally acceptable. The reason is that a different solution would have some undesired consequences and would serve no conservation purpose.

When a coastal state has established the TAC in its zone in accordance with article 61(1) of the LOS Convention, the Convention requires the coastal state to determine what share of this TAC is to be allocated to other states (art. 62). This method of allocation may be applied to exclusive stocks, but in the management of shared stocks the quota allocated to another 'owner' state does not represent the total quota available to that state. The reason is that the same stock may be caught in that state's own zone. In addition to allocating quotas to the other owner state, the owner states must thus also agree upon a sharing of the joint stocks based upon the stocks' whole migration area.

The LOS Convention does not contain any provisions on such sharing. The only provision of relevance is article 63 on the duty to co-operate over the management of such stocks, but this provision gives limited guidance.[29] And customary international law is of even less help on this subject. This means that sharing is essentially a question of negotiation between the states concerned.

There are two main criteria that may be used when sharing joint stocks between the coastal owner states: historic fishing and the distribution of the fish stocks in the zones. Whereas an allocation based on historic fishing would seem to correspond to the former open-access regime, 200-mile EEZs are based on 'ownership' of the fish resources adjacent to the coastal state. It would seem consistent with the philosophy behind the 200-mile system to share the stocks according to the proportion of the stock found in each of the states' zones. It is, however, difficult to establish such proportions as the fish stocks consist of several components: eggs and larvae, young fish and mature fish. In addition, the distribution of fish may depend on the size of the stock.[30]

The North-east Arctic cod was first shared in 1974 on a trilateral basis by an Agreement of 15 March 1974 between Norway, the USSR and the United Kingdom.[31] In 1975 and 1976 quota regulations were adopted by the NEAFC. In the case of the latter, the allocation between Norway and the USSR was basically an equal quota, with some modification. Norway was entitled to over-fish the quota with so-called passive gear, such as net and line. Because Norway's right to a certain extra quota of coastal cod was acknowledged by the NEAFC in 1975, the USSR claimed, and got, a similar quota on Murman cod – which, however, is not recognised by ICES as a stock separate from the North-east Arctic cod.[32] There is every reason to assume that the general principle of equal sharing between Norway and the USSR in this period was based on historic fishing.

After the establishment of 200-mile economic zones, the basic principle of equal sharing of the North-east Arctic cod has been continued. The 'Murman cod' continues to be a part of the sharing and the clause on the 'passive gear' continued until 1982. Norway has, however, been accorded the right to allocate 60 per cent of the total agreed third-state catch while the USSR allocates the remaining 40 per cent. Against this background, it seems that the general principle for sharing cod, developed under the open-access regime, has been carried over to the 200-mile EEZ regime. Haddock has been shared in a similar way to cod.

The sharing of capelin was not discussed by the two states until 1979. As will be seen from Figure 4.3, the USSR rapidly increased its capelin fishing from 1974 onwards, resulting in catches of about 30 per cent of the total catch. The subsequent agreement to share catches with 60 per cent to Norway and 40 to the USSR could be seen as a compromise between historic fishing and the distribution of capelin between the zones.

Unlike the situation regarding the sharing of joint stocks in the North Sea, no comprehensive assessment of stock distribution in the Barents Sea has ever been undertaken. This means that it is impossible to draw any firm conclusions as to the extent to which the sharing of North-east Arctic cod, haddock and capelin conforms to the distribution of these stocks between the two zones. There is, however, sufficient basis for assuming that historic fishing has also played a role when sharing these three stocks. The same allocation has been used for the three stocks every year since the establishment of 200-mile zones. This sharing of the stocks has, however, not been adopted formally, and it is accordingly not

Year	Norway	USSR	Other	Total
1965	217	7	–	224
1966	380	9	–	389
1967	403	6	–	409
1968	522	15	–	537
1969	679	1	–	680
1970	1,301	13	–	1,314
1971	1,371	21		1,392
1972	1,556	37	–	1,593
1973	1,291	45	–	1,336
1974	987	162	–	1,149
1975	943	431	43	1,417
1976	1,949	596	–	2,545
1977	2,116	822	2	2,940
1978	1,122	747	25	1,894
1979	1,109	669	5	1,783
1980	999	641	9	1,649
1981	1,238	721	28	1,987
1982	1,158	596	5	1,759
1983	1,493	846	36	2,375
1984	811	628	42	1,481
1985	453	398	17	868
1986	72	51	–	123
1987	–	–	–	–
1988	–	–	–	–
1989	–	–	–	–

Figure 4.3 International catch of Barents Sea capelin ('000 tonnes) in the years 1965–89

Source: *Reports of the ICES Advisory Committee on Fishery Management, 1989*, International Council for the Exploration of the Sea Co-operative Research Report No. 168, 1990, part I, p. 175.

binding for future years. But politically it would seem that a state seeking to claim a new allocation formula would need very convincing arguments.

We turn now to the allocation of quotas in the 200-mile zone to the other party: article 2 of the 1976 treaty establishes that every year the two parties are, after consultations, to grant access to each other's EEZ. The annual share of the total allowable catch in the zone is to be stipulated according to certain criteria, but these criteria are very vague and not exhaustive, which leaves each state with a wide discretion.

In the yearly protocols adopted by the joint Norwegian–Soviet commission it is recognised that the mutual allocations are based on a balanced exchange of fishing rights. But, again, the problems are somewhat different for shared and exclusive stocks. The allocation of quotas of the coastal state's exclusive stocks determines the other party's total catch of such stocks. Thus, there should be a balance between the mutual allocation of quotas in the two zones. But since both parties have access to the shared resources in their own zone, the allocation in the other party's zone will not necessarily represent the maximum quota available. When considering the balance for shared stocks, it is therefore necessary to assess each party's transfer of part of its share to the other party. The overall balance is then represented by the shares of the exclusive stocks allocated to the other party and the exchange of shares of shared stocks.

There are no agreed principles on how to measure the balanced exchange of fishing rights, and this principle of balance is in any case not legally binding on the parties. Each of the parties may have different economic priorities as to which stocks they prefer. It is consequently difficult to evaluate whether there really is a balance. If prices on Western markets are applied, it seems, however, that the balance is in favour of the USSR.[33]

CONSERVATION AND UTILISATION

The LOS Convention requires coastal states to ensure that fish stocks in the EEZ are not endangered by over-exploitation (art. 61(2)) and that the stocks are maintained at or restored to 'levels which can produce the maximum sustainable yield, as qualified by relevant environmental and economic factors' (art. 61(3)). Coastal states shall also promote the optimum utilisation of the living resources of the EEZ (art. 62(1)) and shall, as stated above, establish the total allowable catch for each fish stock within the EEZ (art. 61(1)).

Because of the language in which these duties are formulated, they allow coastal states wide latitude of action.[34] We would, however, maintain that due to the duty of optimal utilisation, the coastal state may not set the TAC at an arbitrarily low level just to avoid having to give access to foreign fishing vessels. A low TAC must be justifiable according to legitimate concerns (see LOSC, art. 300, on 'good faith' and 'abuse of rights').[35] It is, however, uncertain whether these vague duties have become part of customary international law.[36] And if coastal states disagree on which conservation measures to adopt for shared stocks, these duties are not of such a precision that they render any guidance as to which measures to adopt. For shared stocks, disagreement between owner states may therefore mean that no effective conservation measures are implemented, while none of the states would have violated the LOS Convention.[37]

The Norway–USSR 1976 Treaty is also vague when it comes to management and conservation. Article 2 provides that each party every year shall establish the total allowable catch for separate stocks or combination of stocks, taking into account certain factors. Other conservation measures shall also be established, aiming at conserving fish stocks at a level that can produce the maximum sustainable yield. But both the total allowable catch and other conservation measures shall be established 'as appropriate', which leaves a wide discretion to the coastal state. Moreover, these duties do not provide any guidance as to what kind of management measures to adopt for shared stocks when the two states disagree. Thus, the provisions of the 1976 Treaty on management and conservation are as vague as the LOS Convention and customary international law. This means that the question of which conservation measures to adopt is a political rather than a legal one.

Consultations on conservation between Norway and the USSR in the joint commission have not led to any serious conflicts, so in this sense co-operation has been successful. But one of the main expectations of the new fisheries regime was that it should prevent over-exploitation. Has this happened in the Barents Sea?

The total catch of fish and shrimps in Northern waters increased from about 1 million tonnes in 1947 to approximately 4.5 million tonnes in 1977. Since that year, which was the year 200-mile zones were established, catches steadily decreased to about 1.5 million tonnes in 1986.[38] Since the middle of the 1980s, catches have fallen to one third of the level of catches in the period 1968–88.[39] This development is illustrated in Figure 4.4. Catches have decreased despite increased capacity in the fishing fleet, which means that there must have been a decline in the fish stocks.

The spawning stock of the Norwegian spring-spawning herring used in the 1950s to be about 10 million tonnes, and it was the largest fish resource in Europe.[40] This stock was, however, depleted by heavy fishing in the 1960s, and in 1970 the spawning stock was estimated at a few thousand tonnes. Since then, the stock has been protected by strict regulations and the size of the stock is now at between 1 and 2 million tonnes. Although far from the 1950 level, there has accordingly been a certain increase since the introduction of the 200-mile system,[41] see Figure 4.5. On the other hand, the size of the capelin stock is at a considerably lower level than it used to be and fishing for capelin has been prohibited in recent years. Norwegian and Soviet scientists suggest that fishing for capelin should not be allowed before, at the earliest, 1992.[42] The size of the North-east Arctic cod stock has steadily decreased since 1946 and is now at a very low level (see Figure 4.6 for

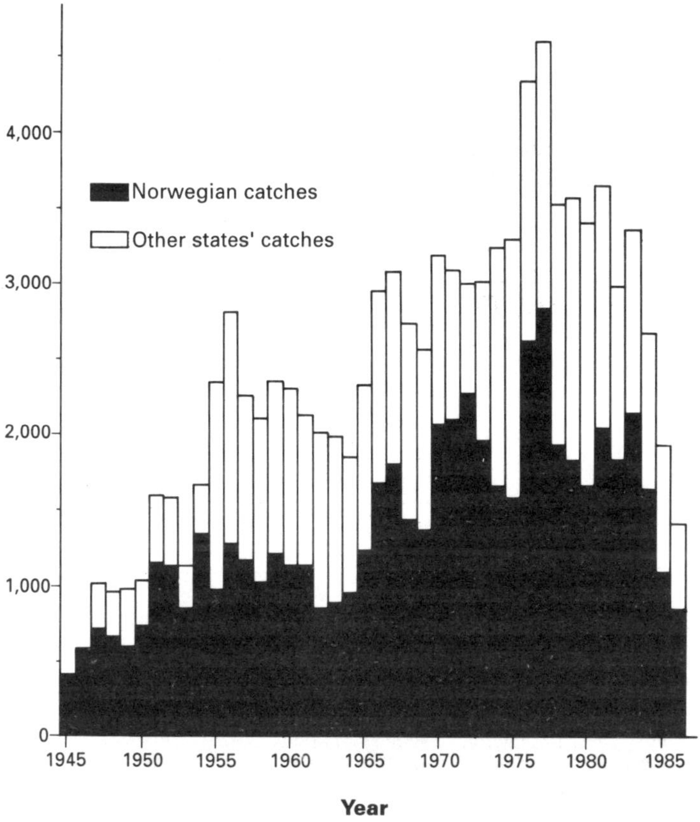

Figure 4.4 Total catch of fish and shrimps in the Barents and Norwegian Seas 1945–86
Source: Norsk Polarforskning, Norwegian Government Publications, Nov. 1989: 9, p. 19

the development of the total stock). The state of the spawning stock may in fact in 1990 have been the lowest in the stock's history.[43]

We may conclude that there has been a positive development in the herring stock in recent years, but that capelin and cod have been depleted. This means that the expectations of the 200-mile zone as regards improved management are far from fulfilled. Why is this so? In order to explain the situation, it is necessary to examine developments both before and after the introduction of 200-mile EEZs.

The state of the fish stocks today can partly be explained by natural fluctuations in the stocks, independent of human activities. A similar

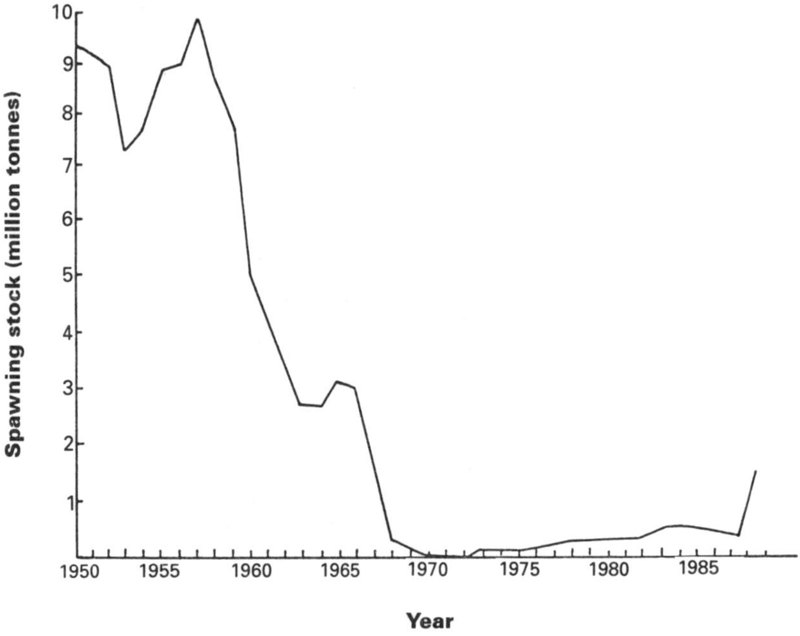

Figure 4.5 Norwegian spring-spawning herring. Spawning stock size in weight 1950–88

Source: *Reports of the ICES Advisory Committee on Fishery Management, 1989*, International Council for the Exploration of the Sea Co-operative Research Report No. 168, 1990, Pt 1, p. 279

crisis in the fisheries in Northern waters is for example described in 1903,[44] which is long before fishing represented a threat to the fish resources. 1976–82 was the longest and most intense period of coldness in the Barents Sea. Such changes of climate probably have great impact on fish recruitment.[45]

But climate changes cannot explain the drastic decline in the herring stock during the 1960s, illustrated in Figure 4.5, and the steady decline of the cod stock since 1945, illustrated in Figure 4.6.[46] Scientists, fishery managers and the fishermen must share the responsibility for this development.

The herring stock was depleted by over-fishing already in the 1960s. This has been called the most negative man-made impact on the ecosystem of the Barents Sea.[47] The decline of the capelin stock occurred after the establisment of EEZs, and is probably a combined result of over-fishing and ecological changes.[48] The over-fishing of

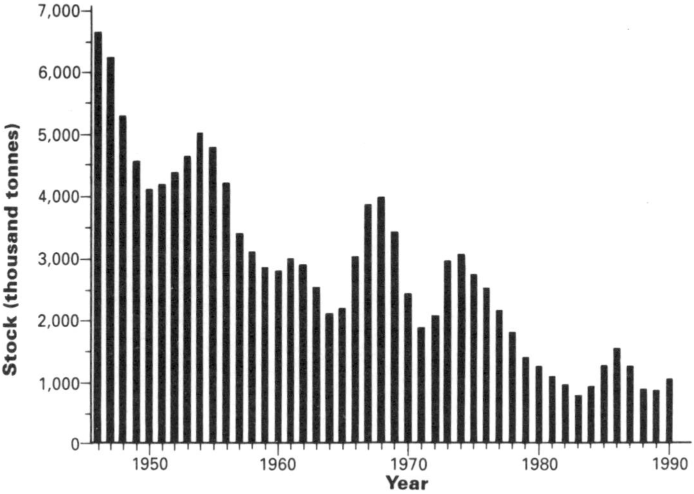

Figure 4.6 The development of the total stock (three years and older) of the North-east Artic cod 1946–90

capelin is indicated in Figure 4.7, which shows that during the eight years 1982–9, there were four years when the TACs adopted were higher than those recommended, and four years when catches were higher than the TACs adopted. Scientists now also claim that the recommended TACs for capelin were probably set at too high a level in 1983–6.[49]

The absence of herring and capelin means that the cod stock has suffered from food shortages, and this has had a negative impact on the cod stock. But there have also been examples of setting the TACs for cod too high in some years and of over-fishing of the TACs set. In the years 1984 and 1985 the TACs set were 47 per cent and 29 per cent in excess of the recommended TACs, while landings of fish were 85 per cent and 81 per cent in excess of the recommended TACs (see Figure 4.8). But the recommended TACs for cod were also probably too high in some years, notably in 1986–8.[50]

There has also been an unsatisfactory catching pattern in the fishing for North-east Arctic cod, with too many small fish being caught. The two coastal states have failed to agree upon the necessary conservation measures. The USSR has traditionally favoured low or zero quotas in this fishery. A zero quota could, however, be a disaster for northern Norway and would make it impossible for Norway to trade quotas in the North with access to the European Community zone. Instead, Norway has advocated the protection of young fish by increasing the mesh size

Year	1982	1983	1984	1985	1986	1987	1988	1989	Max[2]	Min[2]	Mean[2]
Recommended TAC	1600	2300	1100	1000	0	0	0	0	–	–	–
Agreed TAC	1700	2300	1400	1100	120	0	0	0	–	–	–
Actual landings	1759	2375	1481	868	123	0	0	0	2940	0	871
Sp. stock biomass (1 Jan)	582	122	219	83	36	16	167[1]	150[1]	1767	16	454
Recruitment (age 2)	307	425	372	246	11	4	29[1]	187[1]	735	4	386

Figure 4.7 TACs and catches of Barents Sea capelin

1 Predicted or assumed.
2 Over period 1973–89. Weights in '000 tonnes, recruitment in billions.
Source: *Reports of the ICES Advisory Committee on Fishery Management, 1989,* International Council for the Exploration of the Sea Co-operative Research Report No. 168, 1990, Pt 1, p. 29

Year	*1982*	*1983*	*1984*	*1985*	*1986*	*1987*	*1988*	*1989*	*Max*[2]	*Min*[2]	*Mean*[2]
Recommended TAC[3]	<434	<380	150	170	<446	<645	530[4]	300–370	–	–	–
Agreed TAC[3]	300	300	220	220	400	560	590[4]	300	–	–	–
Actual landings[3]	364	290	278	308	430	523	432	–	1,197	278	621
Sp. stock biomass	373	326	281	289	242	262	197	150[1]	677	151	327
Recruitment (age 3)	158	170	389	474	801	199	156	199[1]	1,818	112	451
Mean F (5–10, u)	0.74	0.74	0.88	0.78	0.88	0.95	0.84	–	0.95	0.56	0.75

Figure 4.8 TACs and catches of North-east Arctic cod

[1] Predicted or assumed.
[2] Over period 1969–88.
[3] Norwegian coastal cod not included.
[4] New advice May 1988: 325,000–363,000 tonnes. Agreed TAC reduced to 451,000 tonnes. Weights in '000 tonnes, recruitment in millions.
Source: *Reports of the ICES Advisory Committee on Fishery Management, 1989*, International Council for the Exploration of the Sea Co-operative Research Report No. 168, 1990, Pt 1, p. 9

in trawl fishing and increasing the minimum size of fish caught.

Soviet objections to such measures are probably based on the fact that most of the young fish are in their zone. An increase in the mesh size in trawls and in the minimum size of fish would reduce the fishing opportunities in their own zone. In addition, the Soviet fleet may suffer a greater loss than Norway due to their dependence on trawl fishing.

An increase in the mesh size and the minimum size of fish could accordingly affect the distribution of the resource between the two states. The introduction of such measures calls for compensation from Norway to the USSR to restore the balance in the fishing opportunities for the two states. This illustrates that the concept of balance applies both to the allocation of quotas and to the adoption of other conservation measures.

Norway has contended that one way of compensating the USSR would be to allow more Soviet fishing in the Norwegian EEZ. But this proposition is not very attractive to the USSR because Soviet fishermen at present do not utilise all their quotas in the Norwegian EEZ. They prefer to take the cod in their own zone. The USSR could instead be compensated by Norway accepting Soviet requests, such as making the trawler-free zones smaller and allowing pelagic trawling in the Norwegian EEZ. Transfer payments could also be considered, which have the benefit that the 'partners are encouraged to focus upon the allocation of the net economic benefits from the fishery, rather than upon the allocation of harvests'.[51]

Such compensation may, however, provoke Norwegian fishermen by restricting their fishing. It is worth mentioning that Norwegian fishermen in 1982 protested against the regulations applying to coastal fisheries by continuing to fish when fishing had been prohibited and by blocking a sound in North Norway. These cases even came before the Norwegian Supreme Court, where the fishermen lost their case.[52] Such protests may prevent states from introducing unpopular conservation measures at the national level and may therefore make it even more difficult to reach agreement on an acceptable balance between the two coastal states at the international level.

As a result of the deadlock in negotiations, Norway has unilaterally increased the minimum size for cod, haddock and saithe in the Norwegian mainland zone and the fishery protection zone around Svalbard.[53] The dilemma with such unilateral measures is that they do not apply to the Soviet zone, which contains a high proportion of young fish, or to Soviet vessels in the 'Grey Zone'. Norway may also face problems with enforcing unilateral measures in the 200-mile zone around Svalbard. This means that no satisfactory solution has been found as yet to the need to protect the young cod.

But even proper quotas and other catch regulations will be in vain if they are grossly violated. The magnitude of violations has reached such a level that scientists in the ICES working groups operate with two statistics for catches; the official one – which they do not believe in – and another with their own assessments which is considered to have greater credibility.[54] This emphasises the need for effective enforcement.

Enforcement may, however, present problems which may seem insurmountable. Peter Derham claims that 'from a policing point of view many [regulations] have proved to be so difficult if not impossible to enforce that they seem to have but presentational value'.[55] He concludes quite pessimistically:

> As noted in the early paragraphs of this paper the cost of implementing and enforcing fisheries legislation even at its lowest level can readily absorb the economic benefits likely to accrue from the resource it is designed to protect. Would the answer after all be to forget fisheries management and let economic forces take their toll? [56]

Enforcement may also present another challenge, as illustrated by the claims by Norwegian fishermen that Soviet fishermen by October 1989 had caught 455,000 tonnes of cod while their quota for that year was 134,000 tonnes.[57] The Norwegian Coast Guard claims, however, that the Soviet fishermen comply with regulations in the Norwegian 200-mile zone, while the Norwegian Coast Guard has no authority to control Soviet vessels in the 'Grey Zone' or in the USSR's zone.[58] Such claims by Norwegian fishermen may reduce their willingness to accept strict regulations in fishing for shared stocks. There may therefore be a need to consider 'confidence building' measures to ensure that the regulations are complied with in both zones. This could, for example, be done by giving a mutual right to inspect each other's vessels in the other party's zone or by exchanging inspectors.

A conclusion so far would be that natural fluctuations may change the ecosystem in the Barents Sea fundamentally. Such fluctuations are important factors behind the state of the stocks in this area today. But these fluctuations cannot explain the long-term drastic decline of the herring and cod stocks. This development must be attributed to an ever-increasing capacity and effectiveness in the fishing fleet and unsatisfactory management. This means that there are several problems facing scientists and managers in the years ahead. Some further problems will be added in the following pages, before some general conclusions about the prospects for the future are drawn.

THE ACCESS OF THIRD STATES' FISHING VESSELS

When the TACs for both exclusive and shared stocks have been determined, the next question is to what extent the coastal state is under an obligation to allocate quotas to other states.

Article 62(2) of the LOS Convention provides that the coastal state shall allow other states access to 'the surplus of the allowable catch' (i.e. the balance between the total allowable catch and the coastal state's own catch capacity). Both of these elements are very vague and it is therefore debatable to what extent giving access to the surplus is a meaningful coastal state obligation.[59] We have, however, concluded above that the coastal state may not set the TAC at an arbitrarily low level. Hence, there may legally exist a surplus in certain cases. But to what extent the coastal state is under an obligation to grant other states access under customary international law is very questionable.

If there is a surplus, the next question is who gets what? The LOS Convention deals with these questions in articles 62, 69 and 70. The coastal state is given a broad discretion in deciding which other states' fishermen shall be given access to any surplus. Articles 69 and 70 allow a certain right of access to landlocked and geographically disadvantaged states,[60] but the provisions are fairly tenuous, and the right largely depends on how much a coastal state is prepared to concede when negotiating an access agreement.[61] Under customary international law it would seem that the coastal state is completely free to choose which states are to be granted access.

In the Barents Sea, due to the high catch capacity of Norway and the USSR, there is, however, no surplus of fish stocks of commercial interest. This means that the two coastal states are free to choose to what extent other states should be allowed access to fishing in this area, and if so, which states should have such access.

Norway has, from the establishment of its EEZ, based the allocation of fish quotas in the Barents Sea partly on an exchange of quotas with other states (i.e. the USSR, the European Community (EC) states, and the Faroe Islands), and partly on transitional and surplus agreements with traditional fishing nations in the area (i.e. Poland, the GDR, Portugal and Spain).

The objective of the first type of agreement is to develop a reasonable balance in the reciprocal fisheries. Apart from getting access to the EC's EEZ, it has, however, also been considered to be in Norway's security interests to try to keep some Western states in her EEZ in the North. Such presence reduces some of the unsatisfactory effects of the high degree of bilateral management of the Barents Sea.

The objective of the second type of agreement has been, after a transitional period when traditional fishing was phased out on stocks where Norway had the capacity to take the whole of the allowable catch, to provide access mainly to surplus stocks. The agreements with Spain and Portugal were brought to an end by 1 January 1987, as by that time they had joined the European Community. These two states are now allowed access to the Norwegian EEZ as member states of the European Community. The agreements with the GDR and Poland have been prolonged for a period of five years from 1 January 1987. In 1990 the GDR was allocated minor quotas on redfish (700 tonnes), Greenland halibut (100 tonnes) and blue whiting (7,000 tonnes), whereas Poland has been allowed to fish a small quota of blue whiting (3,000 tonnes). The reason for allocating these quotas is probably not that Norway considers any surplus to exist. The allocations should rather be seen in connection with Norwegian efforts to control fishing in the Svalbard 200-mile zone (see pp. 115–20) as the 'protocols' containing the yearly quotas also contain a clause on the need for limiting fishing in this area. In addition, Norway has probably wished to treat these states favourably, especially considering the recent political changes in these states. But fishermen from the GDR can no longer expect to be allotted special quotas following the establishment of a united Germany as a member of the European Community. As far as the USSR is concerned, it seems that except for granting quotas to Norway to establish the mutual quota balance between the two states, it allocates quotas only to the Faroe Islands.[62]

Third states' share of the catches of cod in the Barents Sea region dropped from about 20 per cent in 1976 to about 10 per cent in 1987 (see Figure 4.9). Of this third state catch about 7 per cent is caught by the enlarged European Community, whereas the Faroe Islands catches about 3 per cent, and the GDR takes only a tiny share. This means that Norway and the USSR only allocate small parts of the cod TAC to third states, and the same applies to other fish stocks. As stated above, the two coastal states do not allocate these quotas because of a legal obligation to allow access to a surplus, but in order either to get access to other states' 200-mile zones or to attain other political goals. Third states' right of access to the Svalbard zone will be discussed below (pp. 116–19).

THE 'GREY ZONE'

The 'Grey Zone Agreement'[63] (already discussed in some degree in Chapter 3) was entered into by Norway and the USSR by an exchange of letters of 11 January 1978. The purpose of the agreement is indicated

Year	Faroes	France	Germany East	Germany West	Norway	Poland	UK	USSR	Others	Total
1976	11,511	20,941	8,946	24,369	344,502	6,986	89,061	343,057	18,090	867,463
1977	9,167	15,414	3,463	12,763	388,982	1,084	86,781	369,876	17,771	905,301
1978	9,092	9,394	3,029	5,434	363,088	566	35,449	267,138	5,525	698,715
1979	6,320	3,046	547	2,513	294,821	15	17,991	105,846	9,439	440,538
1980	9,981	1,705	233	1,921	232,242	3	10,366	115,194	8,789	380,434
						Spain				
1981	12,825	3,106	298	2,228	277,818	14,500	5,262	83,000	–	399,037
1982	11,998	761	302	1,717	287,525	14,515	6,601	40,311	–	363,730
1983	11,106	126	473	1,243	234,000	14,229	5,840	22,975	–	289,992
1984	10,674	11	686	1,010	230,743	8,608	3,663	22,256	–	277,651
1985	13,418	23	1,019	4,395	211,065	7,846	3,335	62,489	4,330	307,920
1986	18,667	591	1,543	10,092	232,096	5,497	7,581	150,541	3,505	430,113
1987	15,036	1	986	7,035	268,004	16,223	10,957	202,314	2,515	523,071
1988[1]	14,478	360	584	2,803	222,844	10,905	8,107	129,600 [2]	1,862	391,543

Figure 4.9 North-east Arctic cod. Nominal catch (tonnes) by countries (Sub-area I and divisions IIa and IIb combined), 1976–88

[1] Provisional figures.
[2] 40,000 tonnes USSR Murman cod not included.

Source: *Reports of the ICES Advisory Committee on Fishery Management, 1989*, International Council for the Exploration of the Sea Co-operative Research Report No. 168, 1990, Pt 1, p. 147

by its title, saying that the agreement is 'interim' and a 'practical arrangement' limited to 'fishing'. The intention was thus to find an arrangement for fisheries management in the disputed area pending a final boundary delimitation. Such an arrangement would make it possible both to control fishing by vessels from the two coastal states and by vessels from third states. In addition, an agreement could prevent conflicts and promote low tension in this area.[64]

The Agreement consists of three parts: the exchanged letters, an annex with the relevant co-ordinates, and a protocol containing substantive provisions and corresponding statements made by the two parties upon the exchange of letters, confirming the parties' intention to reach a final delimitation solution and the non-prejudicial character of the Agreement.

It would have been natural that the geographical application of the Agreement was the same as the disputed area. This was, however, rejected by the USSR, which claimed that the area between the sector line and the median line was not an area of dispute.[65] The result has been that the Agreement geographically is far more favourable to the USSR than an agreement based on the disputed area. This fact created political opposition against the Agreement in Norway, and resulted in the first real split in Norway on a foreign policy issue for a long time.[66]

The letters exchanged provide that co-ordination of the practical implementation of the Agreement, including the setting of the total allowable catch (TAC), shall be executed by the joint Norwegian–Soviet fisheries commission referred to earlier. It is further provided that the TAC shall be utilised in essentially equal shares by the two states, while a part of the TAC is to be allocated to third states after consultation between the parties. Vessels from third states shall be allowed access only if licensed by one or both parties (protocol, art. 4). Conditions for third states fishing are set out (arts 6 and 7) and it is required that the licensing party informs the other party about the vessels licensed (art. 5).

Article 7 of the protocol provides that all fishing shall be conducted in accordance with the regulations on fishing gear, minimum mesh size of trawls and minimum size of fish adopted by the North-east Atlantic Fisheries Commission. In addition, this article establishes a prohibition against third-state fishing of Atlanto-Scandic herring and provides that such fishing by the two parties shall be agreed upon within the joint fisheries commission (litra a).[67] Salmon fishing is prohibited (litra b), and prohibitions against fishing and keeping on board Kamchatka crab (litra c) are provided. These provisions do not apply to fishing for marine research approved by the two parties (art. 9).

Article 8 establishes that the joint fisheries commission may both amend the provisions of articles 6 and 7 and adopt supplementary conditions for third-state fishing. As article 7 stipulates that all fishing activity shall be conducted according to its provisions, it could be claimed that the parties are excluded from adopting different fishery regulations for their own vessels in this area on a unilateral basis. The letters exchanged strengthen such a view by stating that the parties assume the necessity of common regulations for fishing vessels from the two states and from third states. It is clear that one of the parties may not adopt unilateral regulations for vessels from the other party or third states' vessels licensed by the other party. There is, however, no reason why one of the parties should be excluded from adopting stricter regulations for its own fishing vessels and such third states' vessels as it has licensed.

It was important for Norway to ensure that fisheries management in this area was not to be a condominium between the two states. While it has been necessary to establish a joint framework for the adoption of fishing regulations, there is a functional division of enforcement responsibilities. Article 3 prohibits one party from enforcing fishing regulations against vessels from the other party. Both parties may take enforcement action against vessels which do not have a licence from either of the parties (art. 10(b)). Article 10(a) provides that each of the parties enforces the regulations laid down in articles 6 and 7 against third states' vessels licensed by that party.[68] The same must apply to regulations adopted by the joint fisheries commission in accordance with article 8. It must also follow that one party is excluded from enforcing the regulations against vessels licensed by the other party.

The 'Grey Zone Agreement' has been a success from a fisheries point of view in that it establishes the legal framework for the two states' fisheries management in this area. There have not been any serious confrontations over the interpretation or implementation of the Agreement. The two states have also been able to reach agreement on management measures in this area to the same extent that measures have been agreed upon in the rest of the migration area for shared stocks. The Agreement prevents, however, one party from imposing conservation measures on the other party or on third states' vessels licensed by the other party. This means, for example, that the increased minimum sizes of fish, established unilaterally by Norway, are not applicable to Soviet vessels and third states' vessels licensed by the USSR in the 'Grey Zone'. Thus, the 'Grey Zone Agreement' underlines the need for common conservation measures in the Barents Sea.

THE SVALBARD ZONE

As stated in Chapter 2, the 200-mile zone around Svalbard contains important nursery and fishing grounds for such fish stocks as cod, haddock and capelin. Historically, 25 per cent of the cod catches in the Barents Sea have been taken in this area and it is now estimated that 18 per cent of the total Norwegian fish catch is taken in the Svalbard zone.[69]

It will also be recalled from Chapter 2 that there is disagreement between states as to whether the 1920 Svalbard Treaty applies in the 200-mile zone around Svalbard. Norway claims that the Treaty does not apply in this zone, whereas the USSR contends that the Treaty prevents Norway from establishing measures beyond the territorial sea on a unilateral basis. Other states (except for Finland which supports Norway's position) have either claimed that the Treaty applies in the Svalbard 200-mile zone or have reserved their position. The main effect for fisheries of the Treaty applying in the Svalbard 200-mile zone would be a prohibition against Norway adopting discriminatory regulations (cf. the Svalbard Treaty, art. 2).

The 200-mile fishery protection zone around Svalbard was established by Norway by Royal Decree on 3 June 1977,[70] with effect from 15 June the same year. The zone was established on the basis of the Norwegian Law on the Economic Zone of 17 December 1976, but article 2 of the Royal Decree provides that article 3 of the Law, which prohibits foreign fishing, will not 'for the time being' be applied in the fishery protection zone. Norway wanted in this way to signal that it has the right to establish an EEZ around Svalbard, but has not as yet made use of this right. In the meantime, a fishery protection zone based on non-discrimination is being applied. The intention in establishing such a zone was to control fishing in the area without confrontation with other states.[71]

According to the Royal Decree, the Norwegian government may establish total allowable catches, closed areas, minimum mesh sizes and minimum sizes of fish (art. 3). In the Decree it is also stipulated that fishing vessels are to report the commencement and end of their fishing and the size of their catch. They shall also keep a catch log-book (art. 4).

Soviet fishing vessels are not reporting their activities in the zone. Reporting to the Norwegian authorities would be to acknowledge Norwegian jurisdiction in the zone. These difficulties with the reporting requirement are also linked to the unsolved boundary delimitation between Norway and the USSR. The USSR will not report to the Norwegian authorities from areas east of the sector line.[72] Soviet catch data is, however, provided in the bilateral fisheries commission, which means that the USSR's violations of the reporting requirements do not undermine fisheries management in the area.

Regulations on fishing gear, minimum size of fish and closed areas for all fishing, except shrimp trawling, were established on 28 April 1978, following consultations with other states.[73] An area was also closed for shrimp trawling on 5 December 1985.[74] In accordance with an understanding with the USSR, the minimum mesh size in trawl was increased to 125 mm on 8 April 1980. The USSR rejected a further increase in the minimum mesh size, but Norway nevertheless adopted regulations requiring 135 mm on 22 December 1982.[75] It seems that Soviet vessels adhere to these new regulations.

There has also been a need to limit the total catch in the Svalbard zone. Norway and the USSR agree every year on the TAC for cod in the whole migration area, including the Svalbard 200-mile zone. This TAC is shared between the two states and represents the total quota available to each of the states. Thus, it has not been considered necessary to stipulate a quota for each of the parties fishing in the Svalbard zone.

Since there is no third states' quota for cod for the whole migration area, it has been necessary to limit third states' fishing in the Svalbard zone. Norway has tried to keep third states' fishing in the Svalbard zone at a moderate level by allocating quotas in the Norwegian mainland zone on the condition that these states limit their fishing in the Svalbard zone. But as third states' fishing in the mainland zone has been reduced substantially, they do not have the same incentive as before to abide by such a condition.

From 1986 on, Norway has therefore annually established a quota for third states' cod fishing in the Svalbard zone. Third states, such as Spain and Greenland, and the European Community, have protested against these quota regulations.[76] These regulations may be contested as discriminatory on two bases. First, it may be claimed that the quota is set too low. The effect of a too low Svalbard quota would be that more cod would be available to Norway and the USSR in their mainland zones, and thus have a discriminatory effect. The cod quota for third states has been established at 4 per cent of the TAC. It is claimed by Norway that this third states' quota in the Svalbard zone is roughly in accordance with their share of traditional fishing in the area.[77] In addition, the allocation of this 4 per cent is not part of the balance in relation to the mutual access to 200-mile zones in the negotiations between Norway, the European Community and the Faroe Islands. This implies that this share of the cod is allocated to third states on the basis of these states' right to a quota, without any compensation being required. This method of establishing the TAC should therefore probably be accepted as non-discriminatory.

The second basis for contesting these quota regulations would be the way the regulations are designed. Article 1 of the 1990 regulation[78] establishes a global quota of 6,400 tonnes of cod in the fishery protection zone in 1990. But article 2 provides that this quota does not apply to fishing vessels with access to fishing in the Norwegian mainland zone. As long as they are allowed to fish in the mainland zone, fishing in the Svalbard zone may also continue. This means that the quota established in article 1 is without effect for these vessels. It could be claimed by states not having access to the Norwegian mainland zone that this method of regulation is discriminatory.

At the outset, it should be considered as discriminatory to establish quotas which have effect for some states, but not for others. Upon further analysis one will see, however, that the states which will have their fishing limited by the quota would not gain anything if the quota also applied to other states. The result of the exception is that the latter states are allowed to choose whether they would like to catch their cod quota in the Norwegian mainland zone or in the Svalbard zone. Thus, deleting the exception would only disturb these states' fishing patterns and not benefit other states. The design of the quota regulation should therefore be accepted as being non-discriminatory in accordance with article 2 of the Svalbard Treaty. But serious incidents involving violations of Norwegian regulations occurred in 1986, 1987 and 1988. Spanish fishermen over-fished the quota for third states by reporting too much catch on board when entering the zone and too little when they were inside the zone.[79]

So far, violators in the Svalbard zone have only received warnings from the Norwegian Coast Guard; no vessels have been arrested. This means that Soviet vessels regularly receive warnings for violation of the reporting requirements. Spanish fishermen have also received warnings, but the violations in 1988 did not cease until the Norwegian Ministry of Foreign Affairs expressed the seriousness of the situation to the Spanish ambassador.[80] In 1989, the Norwegian Minister of Defence – on behalf of the Foreign Minister – stated in Parliament that, before fishing started that year, contacts with other states at a high political level had been made to prevent violations.[81] The Norwegian Coast Guard has reported that the behaviour of Spanish vessels in the Svalbard zone in 1989 improved.[82]

What about the future of the Svalbard zone? It is obvious that conservation measures are needed to protect the depleted fish stocks. It seems that the cod quota for third states is now accepted in practice. But conflicts may still arise. The European Community claims, for example, that it traditionally has a right to 3.3 per cent of the Svalbard TAC for cod. There is, however, no formal agreement between the Community and Norway about this figure. Furthermore, Spain, which has more than half of the Community's fishing fleet, has not recognised either the Commissions's power to allocate the Svalbard quota among member states, or the Norwegian power to establish quotas in the Svalbard zone.[83] If the size of the cod stock increases, fishing in the Svalbard zone may be more attractive. The result could be more conflicts over third states' fishing rights.

As yet, no quotas have been established for stocks other than cod. But what about measures for stocks that are not shared with the USSR? The Norwegian Foreign Minister, Mr Stoltenberg, announced in Parliament in 1987 that there seemed to be an increasing pressure on fish stocks other than cod in the Svalbard zone. Quota regulations for other stocks could therefore become necessary.[84] Difficulties may then arise both with the USSR and other states.

There may also be a need for other kinds of conservation measures. An increase in the mesh size of trawls may be claimed by other states to be discriminatory, because other states are more dependent on trawl fishing than Norway. It seems, however, that the same conservation purposes may be attained by increasing the minimum size of fish which may be caught, and Norway has introduced such an increase for cod, haddock and saithe, effective from 1 January 1990.[85] An increase in the minimum size of fish must be regarded as a non-discriminatory measure. The USSR has rejected applying such a measure to its own zone. It seems, however, that the new minimum size is being respected in the Svalbard zone.

An increase in the minimum size of fish may be combined with the closure of fishing areas dominated by young fish. This would also be a non-discriminatory measure. But the political effects of a lenient enforcement of regulations closing areas to fishing may be more negative than for other measures; the reason being that illegal foreign fishing in a closed zone would be more provocative to Norwegian fishermen than the more hidden violations of minimum mesh size or minimum size of fish. Disagreement between Norway and the USSR on the need for the protection of the young fish of shared stocks could also result in the transfer of Soviet fishing for young fish from the Svalbard zone to the USSR zone. This means that mutually agreed conservation measures are to be preferred.

In addition to the question of which conservation measures to adopt, there is reason to ask how long the combination of warnings by the Coast Guard with diplomatic contacts will ensure compliance with the regulations. In its 1988 report, the Norwegian Coast Guard stated that the warnings it had issued had so far had no preventive effect. Soviet non-compliance with the reporting requirements is also a worry, in the sense that the USSR may attain a special status in the Svalbard zone.

If Norway is reluctant to implement and enforce the necessary conservation measures, this could have important negative political repercussions on the credibility of Norway's ability to manage the fisheries, and also on its ability to manage future activities on Svalbard's continental shelf.[86] A more long-term effect of the lack of Norwegian ability to introduce and enforce necessary conservation measures could also be that the Norwegian *legal* claim to sovereign rights in the maritime areas around Svalbard is weakened. But implementing conservation measures creates obvious difficulties, and it has been claimed that the management of the maritime areas around Svalbard involves perhaps the most important and most difficult problems facing Norway. It is also claimed that the basic divergencies concerning the interpretation of the Svalbard Treaty may be brought up by such conflicts.[87] Besides undesired higher tension in this strategically important area, the result may therefore be that the International Court of Justice will be asked to solve the dispute. It seems that Norway will try to avoid such a case coming before the Court.

We have in Chapter 2 concluded that the Svalbard Treaty applies beyond Svalbard's territorial sea. And, as far as we can see, there is no development in the attitudes of other states indicating that they will accept the Norwegian view on the area of application of the Svalbard Treaty (which is the opposite of our conclusion). It thus seems that the time has come for Norway to consider accepting the Treaty's application to the continental shelf and in the 200-mile zone. If the Svalbard Treaty applied in the 200-mile zone around Svalbard, the main effect for fishing would be that conservation measures would have to be non-discriminatory. Our discussion has demonstrated that this requirement would not prevent the adoption of necessary conservation measures. The proportion of the total allowable catch that third states would be entitled to under the Treaty would also be rather small, due to their limited traditional fishing in this area.

If Norway accepted the Svalbard Treaty's application in the 200-mile zone, there would be no reason to fear a case before the International Court, because Norway's right to adopt non-discriminatory measures and to full enforcement, including arrests, would be unquestionable. From a fisheries management point of view, there is consequently good reason to recommend that Norway should accept the Svalbard Treaty's application beyond the territorial sea, and commence full enforcement in Svalbard's 200-mile zone. But other considerations, such as petroleum activity on the Svalbard shelf and security aspects, must also be taken into account. This will be done in Chapter 6.

MARINE MAMMALS

Thirteen species of whales have some connection with the Barents Sea, but some of these species are seldom visitors there. The reason is either that their migration area covers only the fringes of this maritime area or that they have been depleted through heavy catching.[88] The bowhead whale was depleted already during the period 1611–1820 in the areas around Svalbard. The blue whale and sei whale met their fate in the period 1868–1904 due to catching off north Norway. As a result of this depletion, only small-scale whaling for the minke whale has been conducted in the Barents Sea in recent years.[89] The Barents Sea also contains seven species of seal, with the harp seal as the commercially most important.

After 1975, the normal annual catch for Norwegian whalers was about 2,000 animals, but as shown in Figure 4.10, the catch has declined. Until the end of the 1970s, Norwegian sealers caught over 100,000 seals annually,[90] but this catch has also declined. Although the exploitation of marine mammals is now of minor importance to the national economy of Norway and the USSR, the catch has still been important to some coastal communities.

The LOS Convention allows coastal states to limit or prohibit the exploitation of marine mammals, rather than establishing an allowable catch and promoting the objective of optimum utilisation (art. 65). States are to co-operate in the conservation of marine mammals.

Whales are managed principally through the International Whaling Commission (IWC),[91] to which both Norway and the USSR are parties. In 1982 the IWC adopted a prohibition on all commercial whaling for the period 1986–90. Japan, Korea, Peru, Norway and the USSR filed objections and the decision was thus not binding for them.[92] However, these states later stated that they would cease whaling and from 1989 on there has been no commercial whaling. Whaling is now limited to catches for research by a few nations, and aboriginal subsistence whaling.[93]

In addition to the whaling industry's importance to local communities, Norway has argued that the minke whale stock is at a higher level than assumed by the IWC. Recent Norwegian research indicates that the North-east Atlantic stock of this whale consists of about 77,000 animals.[94] It should thus be able to sustain a certain catch. Marine mammals also eat a considerable amount of fish and compete with fish for food. There is no evidence, however, that the size of the seal

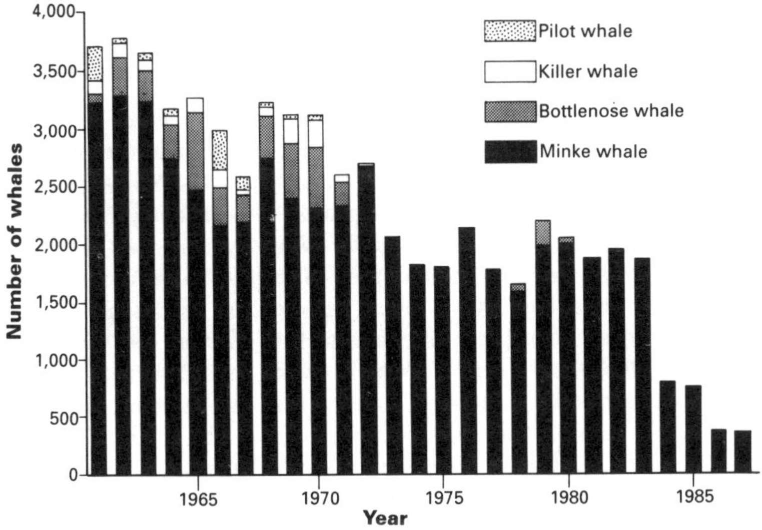

Figure 4.10 Norwegian catches of whales 1961–87

Source: J. Mønnesland *et al.*, *Whaling in Norwegian Waters in the 1980s*, Oslo: Norwegian Institute for Urban and Regional Research, 1990, p. 7

Note: Figures refer to the number of animals

and whale stocks has had a major influence on the decline of fish stocks in the Barents Sea. But it is assumed that harvesting more seals and whales would increase the size of the fish stocks. It has therefore been recommended that 'the sea mammals should be heavily depleted to increase the surplus production of fish resources for man'.[95] Calculations based on simplified assumptions suggest that a catch of 1690 minke whales from the North-east Atlantic stock would make it possible to harvest an extra 6370 tonnes of cod in the first years, which is 5.6 per cent of the total cod quota available in 1990.[96]

The IWC ban was reviewed by the Commission in 1990, but it was not lifted. A resolution condemning the Norwegian research catch was also adopted. The whaling nations find it hard to accept that the IWC obstructs research and is not willing to base its decisions on the best scientific evidence available. The Commission may thus be said to have developed into an organ for conservation, and does not act as a body for both conservation and utilisation. The whaling nations have therefore considered leaving the IWC, but have so far chosen to remain as members.[97] This decision has no doubt been influenced by the US threat to embargo the import of fisheries products under the Pelly Amendment to the Fisheries Conservation and Management Act if whaling continues.[98]

There is no global organisation for the management of seals. As stated above, Norway and the USSR agree every year in the bilateral fisheries commission on mutual catch quotas for seals. Norwegian scientists claim that the number of harp seals is increasing, but Soviet scientists are of the opposite opinion.[99] Anyway, the seals, as competitors in the ecosystem in the Barents Sea, require to be harvested. In recent years there has also been an invasion of harp seals on the Norwegian coast, damaging fishermen's gear; during the winter of 1986/7 about 60,000 seals were trapped in fishing gear along the coast.[100]

The catching of seals does not have to take into account either an international organisation or an embargo from the USA. Environmentalists have, however, campaigned partly against the techniques being used in the catch and partly against any catch at all. The Norwegian authorities have prohibited the catch of baby seals and have appointed expert groups to evaluate the catch methods being used. But the environmentalists' campaign has made it almost impossible to export seal fur. The European Community, for example, has prohibited all import of skins from baby seals since 1983.[101] This means that environmentalists have succeeded both in fighting whaling and sealing. The result is an uncertain future for both these industries, but it also makes an ecosystem approach to ocean management more difficult.

ANADROMOUS SPECIES

Article 66 of the LOS Convention contains special regulations on anadromous species, the most important such species in the North Atlantic being salmon. The state in whose rivers anadromous species spawn is primarily responsible for the management of these stocks. This state may establish the TAC and allow foreign fishermen to fish for any surplus in its EEZ, but it is not obliged to do so. In general, fishing for such species is prohibited beyond the 200-mile EEZ.

Article 6 of the 1976 Norway–USSR Treaty, referred to earlier in this chapter, provides that the state in whose rivers anadromous species have their origin has the main interest and responsibility for their management. The Treaty also provides that fishing for such species should not be conducted outside their 200-mile zones. As stated above, article 7(b) of the protocol of the 'Grey Zone Agreement' prohibits salmon fishing in the area covered by the agreement. The 1982 Convention for the Conservation of Salmon in the North Atlantic Ocean,[102] to which both Norway and the USSR are parties, prohibits fishing for salmon on the high seas, and even prohibits fishing for salmon in most areas, including the Barents Sea, beyond twelve miles. It establishes a North Atlantic Salmon Conservation Organization for co-operation on the conservation and management of North Atlantic salmon. Although there are domestic conflicts between fishermen over access to salmon fishing, the management of salmon does not, however, seem to create international problems in the Barents Sea.

CONCLUSIONS

In the light of past reports on the state of the fish stocks in the Barents Sea, it is obvious that large natural fluctuations in the stocks may occur. The long-term decline in the North-east Arctic cod and herring stock in the Barents Sea cannot, however, be attributed to natural fluctuations, but must be caused by over-fishing. For the most important fish stock, the North-east Arctic cod, this development has continued even after the establishment of 200-mile zones.

This general trend of over-fishing must be attributed to imperfect scientific advice, a lack of ability and will among fisheries managers to follow scientific advice by adopting proper conservation measures, and illegal fishing by the fishermen. The decline in the fish stocks is of special importance for the population of northern Norway, whose livelihood to a large extent is based on fishing. To what extent is an improved management of the fishery resources to be expected?

Scientists still know too little about life in the oceans. Knowledge about the physical and biological phenomena may increase, but the complex ecosystems in the ocean will continue to bring surprises. The 200-mile system has, however, made it easier to adopt adequate management measures in the Barents Sea. Before the introduction of the new ocean regime, all states fishing in this area had to agree in the regional fisheries organisation, the NEAFC, on which conservation measures to implement, while management is now left to the two coastal states, Norway and the USSR (except for whales and salmon).

These two states have sometimes allowed over-fishing by adopting quotas higher than those recommended by scientists. The quota regulations adopted for cod and capelin in the Barents Sea in recent years indicate, however, a willingness to adhere to scientific advice. The

TAC for cod agreed upon by Norway and the USSR for 1990 was actually 12,000 tonnes lower than recommended by ICES. This is promising for future management. But to what extent this trend will continue depends on the two governments' will and ability to let long-term interests prevail over short-term economic pressure.

Norway and the USSR have not been able to agree on which measures to adopt to protect young cod. But this is the only persisting problem that can be attributed to the difficulties involved in the management of shared stocks under the 200-mile system. This problem illustrates the need for agreed measures instead of a unilateral approach. We have, however, pointed out that it should be possible to reach an agreement on the protection of young cod if Norway is willing to compensate the USSR for the disadvantages it will suffer from such measures.

Stricter enforcement is also needed. We have advocated mutually agreed enforcement measures that could be confidence building between fishermen from the two states. Security considerations will, however, prevail over fisheries interests, and as long as Norway considers such measures to involve too much of a condominium between Norway and the USSR, the prospects for mutual enforcement arrangements are bleak.

Delimitation of the continental shelves and the 200-mile zones in the Barents Sea and the status of the 200-mile fishery protection zone around Svalbard are the only legal conflicts connected with fisheries management in this area today. The other management questions are not of a legal, but a political nature. An interim arrangement for fisheries in the disputed boundary area has been found in the 'Grey Zone Agreement', but a solution to the problems connected with the Svalbard zone has yet to be found. We have advocated that from a fisheries point of view, Norway should accept the Svalbard Treaty's application in the 200-mile zone around Svalbard and commence full enforcement of the fisheries regulations. Other considerations must, however, also be examined before a final conclusion is reached on this question.

A rational management of marine mammals seems to face insurmountable difficulties, partly because of the role of the International Whaling Commission and partly due to restrictions on the export of these products. Norway and the USSR could leave the IWC, but the marketing problems and the embargo threat of the USA would still remain. It does not seem, however, that marine mammals are of such an importance in the ecosystem that a multi-species approach to the management of the living resources of the Barents Sea would be totally undermined by the present situation regarding marine mammals.

Our general conclusion would be that the fishing industry must live with large natural fluctuations in the size of the fish stocks. But, if there is a will in the governments of Norway and the USSR to ensure a sustainable yield from the fishery resources, the international management system will not prevent such an objective from being attained.

5 Offshore petroleum activities

INTRODUCTION

As we saw in Chapter 1, the bed of the Barents Sea is thought to contain sizeable quantities of oil and gas. Exploration for these resources has already begun, although it is not so far very advanced. The Norwegians have conducted seismic surveys and exploratory drilling in the south-west part of the Barents, but no commercial finds of oil or gas have yet been made. The USSR is further advanced. It has done more exploratory drilling, and has so far made one commercial find of gas (in the Shtomanovskaya structure, west of the southern island of Novaya Zemlya), although production from this field is unlikely to begin for several years. In addition, since 1987 commercial production has taken place from an oilfield onshore on Kolguyev Island; it is thought that this field may also extend out under the sea.

The first part of this chapter examines the factors that will determine whether, and to what extent, offshore exploration and exploitation for oil and gas in the Barents Sea will continue. This is followed by an overview of the legal regime that governs and will govern present and future offshore oil and gas activities, including the question of how far it is possible for companies from third states to participate in such activities. The final section of the chapter examines the current law governing pollution from offshore oil and gas activities, and considers whether it is adequate.

FACTORS DETERMINING THE LEVEL OF OFFSHORE ACTIVITIES

As we have just seen, the extent of offshore exploration and exploitation for oil and gas in the Barents Sea is currently not very great. The aim of this section is to examine those factors that are likely to determine the future extent of such offshore activities. Such factors are both positive (in the sense of encouraging a relatively rapid development of offshore activities) and negative (in the sense of constraints which will either slow down or possibly prevent such a development). Each type of factor will be considered in turn.

Positive factors

These factors differ as between Norway and the USSR, so that it is necessary to consider each separately. In the case of Norway there appear to be two main factors which would encourage an expansion of its offshore activities in the Barents Sea. The first of these is economic. Since offshore production of oil and gas began in the Norwegian sector of the North Sea in the early 1970s, the offshore petroleum industry has been an important factor in Norway's economic development, both because of its contributions in the form of taxes and licence fees to the Norwegian exchequer (which at their height in 1985, accounted for about 19 per cent of all state revenues)[1] and because of the boost it has given to employment and other sectors of the economy. Since the dramatic drop in oil prices in 1986, the contribution of the offshore petroleum industry to the Norwegian exchequer has fallen drastically, to about 7 per cent.[2] Thus, the possible production of oil and gas in the Barents Sea would represent a new source of tax revenues. More importantly perhaps and in the longer term, oilfields in the North Sea will begin to become exhausted in the late 1990s and in the early part of the next century. There is therefore a great need to find alternatives, not merely to continue tax revenues but also to maintain the general stimulus to the Norwegian economy that the offshore petroleum industry has given it. Outside the North Sea, the Barents Sea is one of the most promising areas of the Norwegian continental shelf for finding such an alternative.

The other factor which would encourage Norway to develop its offshore activities in the Barents Sea is one of regional policy. It has been the policy of successive Norwegian Governments for many years to take measures to try to maintain the level of population in northern Norway (not least for strategic reasons) against the tendency for the population to drift southwards. An offshore petroleum industry centred on the Barents could give an economic boost to northern Norway, in particular by creating employment and work for firms in the offshore supply industry.

We turn now from Norway to the USSR. Up to about 1988 Western analysts[3] identified a number of factors which were likely to encourage the USSR to develop its offshore industry in the Barents Sea. Like Norway, the USSR needs to find alternatives to replace its existing oilfields (mainly on land) which are beginning to become exhausted. The reasons for such a need are not quite the same as with Norway, however. While in the USSR, as in Norway, the oil industry gives an enormous boost to the economy, in the USSR oil is a major source of energy (unlike in Norway, whose energy needs are almost entirely provided by hydroelectric power). Above all, however, as the leading export of the USSR, oil is its major source of foreign currency. In addition, oil also has an important foreign policy dimension for the USSR. Being far more than self-sufficient in oil has allowed the USSR to remain relatively detached from much of the political turmoil of the Middle East. Taking all these factors together, there are thus very good reasons why the USSR needs new sources of oil, particularly as there are few immediate alternatives to oil both as a source of energy and as a major export. Of the various potential new sources of oil, both onshore and offshore in the USSR, the Barents Sea is regarded as one of the more (although not among the most) promising. Official recognition of this is provided by the fact that in 1986 a new Five Year Plan for the first time called for the development of offshore petroleum exploration.[4] As was said above, the situation just described was the perception of Western analysts up to about 1988. Although there do not appear to have been changes as such in these positive factors which would encourage the development of the offshore petroleum industry in the Barents Sea, there must now be some question as to whether they could operate in the way suggested, in view of the deteriorating economy and increasing political instability of the USSR. In other words, even if there was the desire on the part of Soviet decision-makers to develop offshore resources, it might be that economic and political factors would prevent this happening.

Having looked at the positive factors relating to the possible development of the offshore petroleum industry in the Barents Sea, we must now turn to look at the negative factors.

Negative factors

Although we have identified a number of factors that suggest that both Norway and the USSR would be likely to wish to increase the scale of their present offshore petroleum activities in the Barents Sea in the future, there are a number of constraints that might prevent such a wish being realised, or at least slow down the rate of any future development. These constraints are of various kinds – technical, economic, environmental, security and legal.

As far as technical constraints are concerned, exploration and exploitation of the Barents Sea for oil and gas offer a number of challenges to current offshore technology. Many of the more promising areas lie in depths of water which are close to the limits of current technology. In addition, much of the Barents is, as we saw in Chapter 1, covered by ice for part of the year: the technology to drill in ice-covered areas is only now being developed. These problems are exacerbated by the frequent bad weather that occurs in the Barents and the fact that for several months of the year the area is in near total darkness. Even in the southern Barents where the technological problems are not so great, there is a particular difficulty for the USSR in that it lacks experience in the use of offshore petroleum technology – and indeed lacks some of the technology itself – and it has been widely suggested that the work it has so far carried out in drilling offshore in the Barents has, by Western standards, been poorly executed.[5] A final technical obstacle to offshore development in the Barents Sea is the lack of the necessary infrastructure in the adjoining land regions and the distance of these regions from the main markets for any oil or gas found. For example, because the nearest demand for gas is in Germany and no gas pipeline exists down through Scandinavia, then even if gas were found in the Norwegian sector of the Barents it is unlikely that there would be any commercial production, at least in the foreseeable future.

Turning from technical to economic constraints: while price and profitability will probably have limited influence on any Soviet decision, they are crucial as to how far and fast developments take place in the Norwegian sector. Drilling in the Barents is more expensive than in most offshore areas: it has been calculated, for example, that the cost of drilling a well in the Barents Sea is about $27 million, as against $15–20 million in the northern North Sea.[6] Thus, drilling in the Barents depends on the price of oil being fairly high. Predicting the future price of oil is always difficult, but probably impossible at the present time in view of the crisis in the Gulf. It is also worth bearing in mind that if states agree to reduce emissions of greenhouse gases in order to try to slow down global warming, this will lead to a reduction in the demand for oil. On the other hand, in the long term, as world supplies of oil gradually dwindle, the price of oil will, other things being equal, rise. Even if the USSR's actions are not greatly affected by the price of oil, it does nevertheless need money to pay for any development of its offshore industry in the Barents. The USSR is currently short of capital for large-scale projects, and thus is unlikely to be able to go ahead in the Barents unless it has an infusion of Western finance.

Another kind of constraint consists of environmental factors. Such factors are likely to weigh more heavily with Norway, being much more environmentally conscious, than with the USSR. The Norwegian authorities are clearly determined to keep the risk of any accidental large-scale pollution resulting from offshore oil exploration and production to the absolute minimum. It is very important that, if at all possible, there should be no such pollution. In cold sub-Arctic waters, like those of the Barents, oil is broken down much more slowly by bacteria than happens in warmer waters. Fish eggs and larvae that come into contact with oil are killed or deformed; mature fish may become ill or deformed; and the plankton on which fish feed are also damaged or killed. As pointed out in Chapter 1, the Barents is a major spawning and nursery ground for a variety of fish stocks: a major oil spill could therefore have a very damaging effect on the fisheries of the Barents and the livelihoods of fishermen.

The risk of pollution is not the only reason why fishermen would be – and indeed are – concerned about the development of the offshore petroleum industry in the Barents Sea. Other concerns include loss of access to fishing grounds as a result of the presence of offshore oil installations; damage that may be caused to fishing gear and vessels by debris from the offshore oil industry; and the obstacles to navigation between ports and fishing grounds posed by offshore installations.[7] These concerns of fishermen are likely to be regarded as a constraint on the further development of the offshore oil industry in the Barents, at least by the Norwegian authorities, though not perhaps by the Soviet authorities.

Military factors will also play a constraining role. As we saw in Chapter 3, offshore installations can obstruct the passage of naval vessels, and may also have some limited potential for espionage. Because of its sensitivity to Soviet security interests, Norway may feel constrained not to allow emplacement of installations in its sector of the Barents to become unduly obstructive to the passage of Soviet naval vessels. The USSR, for its part, will also be concerned not to let the presence of installations in its sector pose undue restrictions on the activities of its naval vessels.

The final kinds of constraint are legal ones. As we saw in Chapter 3, as long as there is no agreement between Norway and the USSR on a continental shelf boundary, there is an obligation on Norway and the USSR not to engage in exploratory drilling and production in the disputed area lying between the median and sector lines: in other words, there is currently a sizeable area of the sea-bed of the Barents Sea that is out-of-bounds to offshore petroleum development. Second, as long as the question of whether the Svalbard Treaty applies to the maritime zones of Svalbard beyond the territorial sea remains unresolved, it seems unlikely that Norway will engage in much more than seismic surveying on Svalbard's continental shelf. Thus, another sizeable area of the sea-bed of the Barents Sea will remain closed to exploratory drilling and production for some time. Finally, there are a number of constraints resulting from general international law (notably the 1958 Convention on the Continental Shelf and the 1982 UN Convention on the Law of the Sea) such as the obligation not to place installations 'where interference may be caused to the use of recognised sealanes essential to international navigation'.[8] Some of these constraints are discussed in the next section.[9]

Thus, there are both factors which suggest the likely development of the offshore oil and gas industry in the Barents Sea in the near future, and factors which work against such a development (including the fact that very limited discoveries of oil have so far been made). Exactly what the outcome will be is really impossible to say, especially as some of the factors, such as the price of oil, and domestic political and economic developments in the USSR, are highly unpredictable.

THE LEGAL REGIME GOVERNING OFFSHORE PETROLEUM EXPLORATION AND PRODUCTION

Although offshore petroleum exploration and production is, as we have seen, not yet very far advanced, there is already a detailed legal regime governing the matter. The aim of this section is to give a brief overview of this regime. A lengthy book could easily be written on this question, but for reasons of space only the barest bones of the regime can be indicated here. The regime has in fact three components – the international law regime, which applies to the whole of the Barents; Norwegian law, which obviously applies only to the Norwegian sector; and Soviet law, which applies to the Soviet sector. Each of these will be discussed in turn.

The international law regime[10]

As we have seen earlier, in Chapters 1 and 3, the whole of the sea-bed in the Barents Sea is, in legal terms, continental shelf, belonging to Norway and the USSR, even though the exact demarcation of their respective continental shelves has not yet been established. We also saw in Chapter 1 that under international law a state has not only the right to explore and exploit the natural resources of its continental shelf, but also has the exclusive right to do so. In other words, no other state can explore and exploit the resources of its continental shelf without its express permission. The way in which a state exercises its continental shelf rights is very largely left by international law to individual states to determine: for example, international law says nothing about how a state should regulate the exploration and exploitation of its continental shelf resources (for instance, whether by licensing or some other system; or whether foreign companies may participate); the kind of safety standards it should prescribe; the kind of royalty and the tax revenues it may charge; the depletion policy it should follow, and so on. All that international law does is to lay down some fairly broad obligations to which the exercise by a state of its continental shelf rights is subject. The main such obligations are as follows.

First, where the status of the waters overlying the continental shelf is exclusive economic zone (EEZ), as is the case in most of the Barents Sea (see Chapter 1), there is an obligation on the coastal state not to interfere unjustifiably with the navigational and other rights that other states enjoy in the EEZ.[11] Where the overlying waters are high seas, as is the case with a small triangular-shaped area in the middle of the Barents Sea (see Map 3.2), the obligation is not to interfere unjustifiably with other uses of the high seas.[12]

Second, although international law expressly gives states the right to erect such installations as are necessary for exploring and exploiting the resources of their continental shelves, to establish 500-metre safety zones around such installations and to exercise their jurisdiction over such installations,[13] it does require states not to place installations 'where interference may be caused to the use of recognised sealanes essential to international navigation'. In addition, to minimise the possibility of further interference with shipping and other uses of the sea, due notice must be given of the construction of installations, and permanent means for giving warning of their presence must be maintained, such as lights and foghorns.[14] Once installations have served their purpose, they must be removed. The precise scope of this obligation is not altogether clear. Under article 5(5) of the 1958 Convention (to which both Norway and the USSR are parties) 'any installations which are abandoned or disused must be *entirely* removed' (emphasis added). The 1982 Convention, on the other hand, contemplates the possibility of only partial removal. Its article 60(3) provides that disused installations 'shall be removed to ensure safety of navigation, taking into account any generally accepted international standards' (such as the guidelines adopted by the International Maritime Organisation in October 1989)[15] and having 'due regard to fishing, the protection of the marine environment and the rights and duties of other states', with appropriate publicity being given to installations only partially removed. In spite of the fact that the 1982 Convention is not in force, the trend of state practice is moving towards the formulation of a new rule of customary law that would in some cases require no more than the partial removal of obsolete installations (though in other cases complete removal would be required). Such a rule would supplant the seemingly unequivocal obligation of the 1958 Convention to remove all obsolete installations entirely. By the time the removal of disused installations in the Barents Sea becomes a practical issue, which will not be for some decades, it is likely that the trend referred to will have been completed, and that only partial removal of those installations not hazardous to navigation or fishing will be required.

Were any production of oil or gas to take place in the triangular-shaped area lying beyond the 200-mile EEZs of Norway and the USSR, and were the 1982 UN Convention to be in force and Norway and the USSR parties to it, there would be an obligation on whichever state(s) from whose continental shelf such production came, to pay to a body known as the International Seabed Authority (which is to be established when the Convention comes into force) a proportion of the value or volume of such production after the first five years of exploitation.[16] Finally, international law imposes certain obligations on states to control pollution which could result from the exercise of their continental shelf rights. These obligations are considered in the final section of this chapter.

Having looked at international law and found that it provides no more than a framework regime for the exploration and exploitation of oil and gas from the sea-bed, we must now turn to Norwegian and Soviet law to see how they provide a more detailed regime within the broad international law framework just described.

The Norwegian legal regime[17]

Ever since it began authorising exploration and exploitation for oil and gas on its continental shelf in 1963, Norway has utilised a licensing system. Under this system areas of sea-bed are divided up into blocks, usually rectangular in shape. From time to time a licensing round is held in which a number of such blocks is made the subject of invitations to petroleum companies to apply for licences to explore and exploit such blocks. Once such applications have been received, licences are awarded by the Ministry of Oil and Energy on a discretionary basis in accordance with criteria which have previously been laid down. By 1990 thirteen such licensing rounds for blocks on the Norwegian continental shelf had been held.

Licences are of two types – exploration licences and production licences. The former, normally awarded for three years, are limited to 'geological, geophysical, geochemical and geotechnical investigations, including shallow boring'.[18] They give no exclusive right to explore the area concerned, nor does the holder of such a licence have priority when it comes to awarding a production licence for the area concerned.[19] A production licence gives its holder the exclusive right to explore for and produce petroleum in the designated block.[20] Production licences can be awarded only to companies formed in accordance with Norwegian law.[21] Thus, foreign oil companies can only apply for production licences through the medium of a Norwegian subsidiary. In practice, Norwegian subsidiaries of foreign companies currently hold about 40 per cent of production licences: for foreign policy and security considerations it may be that the Norwegian authorities would wish to see this percentage reduced in the Barents Sea, especially in areas of the Norwegian continental shelf close to the Soviet sector (although security considerations would also argue for the presence of some subsidiaries of Western oil companies). The holder of a production licence is subject to a wide range of obligations, such as the payment of a licence fee and royalties, obtaining approval for a programme of work, complying with the rate prescribed at which he is to exploit his block, obtaining permission for the laying of pipelines, and so on.

Under the 1985 Petroleum Law (s.7), unlike the earlier legislation which it replaced, before new areas can be licensed for exploratory drilling or production an impact analysis must be prepared. This evaluates the different interests in the area concerned and the effect of potential oil activity on other industries and the environment. Interested organisations must be informed when such an analysis has been made and given a period of not less than three months in which to express their views. On the basis of the impact analysis and the subsequent consultation a decision on whether to open the area concerned to offshore petroleum activities is then taken by the Government. In the case of the Barents Sea such an impact analysis was completed in 1988 for part of the south-western area of the Barents Sea and an adjoining area outside the Barents, an area known collectively as Barents Sea South, comprising about 210,000 square kilometres, about half of which is outside the Barents Sea proper (see Map 5.1). On the basis of this impact analysis and subsequent consultations, the Government decided in March 1989 that, because of the need to find more oil and maintain levels of investment on the Norwegian continental shelf, and given that

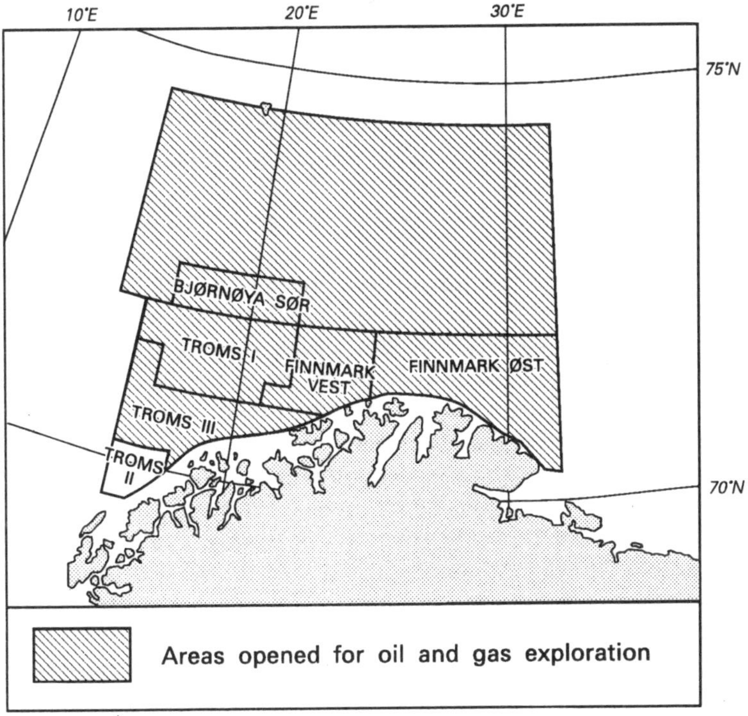

Map 5.1 Barents Sea South

the risk of pollution and of harm to the marine environment from exploration activities was very small, the whole of the Barents Sea South, except for a small area off the coast of Troms (Troms II) (i.e. outside the Barents Sea) should be opened up for exploration, though limitations on the rate and times of the year at which this should be done should be imposed.[22] As a result, twenty-five blocks in the Barents Sea were put on offer and awarded in the thirteenth licensing round held in 1990.[23]

Should the result of the licensing just referred to lead to discoveries of oil or gas, and permission for production to begin be sought and given, then the licence holder would be subject to the variety of obligations outlined briefly above. In addition, the licence holder would be subject to whatever taxation regime were in force at the time, safety regulations,[24] and various obligations designed to prevent pollution (which are discussed in the final section of this chapter).

Finally, given the impact (explained earlier) that any appreciable level of offshore oil and gas activity would have on the fishing industry in the Barents Sea, it is worthwhile considering briefly the main provisions Norwegian law makes for ameliorating the disadvantages for the fishing industry of offshore oil and gas activities.[25] First, fisheries interests are one of the factors to be taken into account in preparing an impact analysis of the kind referred to above. Second, before the holder of a production licence can erect an installation, he must state the effects of the proposed emplacement of the installation on other activities, and he can be required to prepare or pay for an impact analysis on this matter. In the light of this information and any impact analysis, the Department of Oil and Energy then decides whether to give its consent to the proposed emplacement.[26] This would, therefore, in theory at least, allow the Department to prevent the erection of installations in areas that were particularly important to the fishing industry. Third, in other areas of the Norwegian continental shelf, the Norwegian authorities in practice consult with fishermen's organisations over the course of pipelines before giving permission for pipelines to be laid, although this is not required by law. It may be expected that this practice will continue in the Barents. Fourth, under amendments made to the Petroleum Law in June 1989,[27] a fisherman who has suffered economic loss where occupation of fishing grounds by offshore petroleum activities has made it impossible or significantly more difficult for him to fish may claim compensation from the state. Fifth, regulations of 1979[28] prohibit the dumping of debris from the offshore oil industry in the sea (something which has been a particular problem for fishermen in the North Sea), and licensees can be required to remove any debris which has been dumped in breach of the regulations. Lastly, under the 1989 amendments to the Petroleum Law just referred to, a licensee has strict liability in respect of all damage caused to fishermen by debris. Where the origin of the debris cannot be identified, all licensees are jointly liable.

The Soviet legal regime

As a centrally state-planned and -owned economy (at least at present), the USSR obviously has a rather different offshore petroleum exploration and production regime from that of Norway. Under Resolution No. 564 of 18 July 1969 of the Council of Ministers,[29] in order to explore and exploit the natural resources of the USSR's continental shelf it is necessary to register with the Soviet Ministry of Geology (in the case of exploration) and with the State Mining Inspectorate (in the case of exploitation).[30] The erection of installations requires a permit from the

Ministry of Defence. Detailed regulations governing the procedures for registration and obtaining a permit are laid down by the various Soviet authorities just mentioned. While the 1969 Resolution may suggest other possibilities, exploration and exploitation of continental shelf resources are in practice carried out exclusively and directly by Soviet Ministries themselves, especially the Ministry of Gas, and by specialised agencies within these ministries, such as the Arktikmorneftegazrazvedka Trust, which is responsible for conducting exploratory work in the Barents Sea.[31]

Both the Decree of the Presidium of the Supreme Soviet of the USSR of 6 February 1968[32] and the Decision of the Presidium of 13 August 1969[33] contemplate the possibility of foreign companies being allowed to explore and exploit the USSR's continental shelf, either by means of an agreement between the USSR and the foreign state concerned or by means of special permission granted by the competent authorities of the USSR. In January 1987 legislation was adopted which enables equity joint enterprises to be formed in the USSR with foreign partners, and in May 1989 a decree was adopted which encourages Soviet enterprises to form offshore companies in co-operation with foreign companies.[34] The latter is a concrete expression of the call by Mr Gorbachev in his Murmansk speech of October 1987 for co-operation in the development of Arctic resources. So far, limited use appears to have been made of these new legislative provisions as far as offshore petroleum activities are concerned.[35] A joint venture agreement has been proposed between the Soviet Ministry of Gas and a group that includes the Norwegian company, Norsk Hydro, as well as American and Finnish interests, to study the point at which development of the giant Shtomanovskaya gas field can begin.[36] A level of involvement lower than that of joint ventures by foreign companies in the Soviet offshore petroleum industry has, however, been quite widespread for a number of years. Thus, for example, technical co-operation agreements have been concluded between the USSR and a number of Western companies; Finnish companies have built several drilling vessels for the USSR; and French and Norwegian companies have supplied various kinds of technical equipment.[37] The reasons why this account so far has put so much stress on the possible role of foreign companies in the exploration and exploitation of the USSR's continental shelf are partly the anticipated predominantly non-Soviet readership of this book, and partly that a number of Western observers[38] believe that without the input of Western technology, expertise and possibly capital, the USSR will not be in a position satisfactorily to exploit the resources of its continental shelf.

Once the necessary registrations and permissions have been obtained, activities on the Soviet continental shelf, whether by Soviet bodies acting on their own or in co-operation with foreign companies, must be carried out in accordance with the relevant legislation in force, in particular the regulations concerning the construction and operation of installations and those concerning safety.[39] As far as the authors have been able to discover, the USSR (unlike Norway) has no particular legislative provisions designed to try to reduce possible conflicts between the fishing and offshore petroleum industries. Trying to resolve such conflicts would appear to be a matter solely of administrative discretion for the Soviet authorities. The one qualification would be in respect of those few non-Soviet vessels permitted to fish in the USSR's EEZ: in respect of these the USSR is under an international legal obligation to conduct its offshore oil and gas activities so as not to thwart those rights to fish which have been accorded to non-Soviet vessels.

An assessment of the legal regime in terms of marine management

From the point of view of marine management, an assessment of the legal regime just described depends on what one expects the role of management in relation to the exploitation of sea-bed mineral resources to be. The authors would suggest that a proper system of management would have three primary functions. These are, first, to provide for the orderly exploitation of the resources; second, to seek to minimise conflicts between the exploration for and exploitation of these resources and other uses of the sea; and third, to prevent if possible, but if not to keep to a minimum, pollution and other harm to the marine environment from the exploitation of sea-bed mineral resources.

As far as the first of these functions, the orderly exploitation of resources, is concerned, the Norwegian legal regime seems quite satisfactory on paper. So far (which is not far), it has functioned quite satisfactorily in practice, and experience with the operation in practice of the Norwegian legal regime in the North Sea suggests its further practical application in the Barents Sea is likely to be satisfactory. The Soviet legal regime also appears satisfactory on paper, but there must be some uncertainty as to how well it will function in practice in the Barents Sea. This uncertainty is due, first, to the reasonably well-documented allegations of poor workmanship that have already occurred in offshore operations in the Barents, which were mentioned earlier; and, second, to the fact that the USSR is also alleged to have wasted much oil, been inefficient and damaged the environment in its oil exploration activities on land in Siberia.[40] As long as the dispute over the application of the Svalbard Treaty offshore remains unresolved, a possibility exists of a less than orderly exploitation of any oil or gas that there may be in Svalbard's continental shelf, particularly if Norway should decide unilaterally to begin offshore oil and gas activities. If the dispute is resolved, and the solution is that the Svalbard Treaty and the Mining Code do apply to Svalbard's continental shelf, there may be some doubt as to whether the exploitation of any oil or gas by states parties to the Treaty will be sufficiently orderly, in view of the fact that the Mining Code is not well suited to regulate offshore gas and oil activities – a matter which was discussed in the latter part of Chapter 2. On the other hand, the unresolved continental shelf boundary dispute does not pose any likelihood of a less than orderly exploitation of resources as long as both sides continue to respect their obligations not to drill in the disputed area.

We turn now to the second function of a sea-bed resource management regime: to minimise conflicts between the exploitation of sea-bed resources and other uses of the sea. As far as the Barents Sea is concerned there are two uses with which sea-bed resource exploitation may in practice conflict: these are shipping and fishing. As far as the former is concerned, the various international law obligations, which were discussed earlier relating to the siting of installations, should be sufficient to prevent any serious conflicts between shipping and sea-bed resource exploitation – at least as far as civilian shipping is concerned. The position with military shipping is perhaps more problematical. As regards possible conflicts with fishing, this is governed primarily by national law. Norwegian law, as we saw, goes to considerable lengths to minimise and reduce the impact of conflicts between fishing and offshore oil and gas exploitation, but if the latter is to take place a certain level of conflict is inevitable – as has been demonstrated, for example, in the North Sea. While Norwegian law deals relatively extensively with the question of the relationship between fishing and offshore petroleum activities, Soviet law appears to be completely silent on this issue, and it remains to be seen how the Soviet authorities, through administrative practice, will seek to reconcile their differing interests in the Barents Sea of fishing and oil and gas exploitation.

The third and final principal function of a sea-bed resource management regime, it was suggested, is to prevent or minimise pollution and other harm to the marine environment from offshore petroleum activities. This matter is the subject of the final section of this chapter.

POLLUTION FROM OFFSHORE PETROLEUM ACTIVITIES [41]

It is most important that pollution from offshore petroleum activities in the Barents Sea should be kept to an absolute minimum and, if possible, be prevented altogether. As was pointed out above, oil is broken down very slowly in sub-Arctic waters, and any large-scale oil spill could have a very damaging effect on fish stocks in the Barents Sea. It is right that concern about possible pollution should be expressed at this early stage in offshore petroleum activities in the Barents Sea, because of offshore experience elsewhere in Norway and the USSR. In the North Sea, where Norway is generally thought to have a good environmental record, 3,256 tonnes of oil on average were being discharged annually into the sea from petroleum activities on the Norwegian continental shelf during the heyday of production in the mid-1980s.[42] Earlier, in 1977, there was a blow-out in an oil well in the Ekofisk field, when 30,000 tonnes of oil were discharged into the sea. As far as the Soviet Union is concerned, its record also gives cause for concern. Thus, for example, a blow-out at an offshore gas well in the Caspian Sea lasted for eight months before it was brought under control.[43]

When pollution from offshore petroleum activities does occur, it is usually at the production phase. Although no production of oil and gas in the Barents Sea is yet taking place, it is worthwhile considering what law, both national and international, currently exists to minimise pollution and what further laws (or other measures) might be required. That is the aim of this section of this chapter.

Before considering the existing legal regime relating to offshore pollution prevention in the Barents Sea, it is necessary to discuss briefly the ways in which pollution from offshore petroleum activities can occur. Such pollution can be either deliberate or accidental. Deliberate pollution includes the following: the discharge of drilling-muds and cuttings containing oil; the discharge of production water containing oil;[44] the drainage of oil from installations into the sea; the discharge of chemicals used in drilling; and the discharge of debris, other garbage and sewage from installations and supply vessels. Accidental forms of pollution include: blow-outs (i.e. the escape of oil or gas resulting from the loss of control over the flow from a well); rupture of a pipeline; a collision between a ship and an installation; and, where ships are used instead of pipelines to bring oil ashore, the accidental spillage of oil while loading the ship (e.g. as a result of the failure of hoses and couplings).

Having identified the main ways in which pollution from offshore oil and gas operations may occur, we will now briefly examine the international and/or national law relevant to each. To economise on space, where a matter is regulated by international law, no reference will be made to national law which simply implements such international legal provisions.

Drilling-muds and cuttings

This matter is governed at the regional international level by the Paris Convention.[45] The Paris Commission, established by the Convention, in 1986 adopted a decision establishing standards for the use of oil-based drilling-muds.[46] Two years later the Commission adopted a further decision setting stricter standards for the use of such muds.[47] Both of these decisions are legally binding on states members of the Commission – which include Norway – except for those states which have objected to the decisions (which Norway has not done). The area of application of the Convention includes the Barents Sea. However, unlike Norway, the USSR is not a party to the Convention. Thus, the decisions only apply to the Norwegian sector of the Barents Sea. The decisions lay down a number of rules and principles (of whose details lack of space precludes an account) which are designed to avoid and prevent to the greatest possible extent pollution from drilling-muds and drill cuttings. Whether the standards laid down in these decisions ought to be made stricter for the Barents Sea, and whether this is feasible, the authors are not competent to judge. It should be noted, however, that the oil industry is currently developing the use of drilling fluids with no or very little oil.

As mentioned, the USSR is not a party to the Paris Convention, so that the Paris Commission's decisions do not apply to the Soviet sector of the Barents Sea. It would therefore seem desirable that the USSR should become a party to the Convention, as indeed the Norwegian Government has suggested. Whether at the national level the USSR has any regulations laying down standards for the use of drilling-muds the authors have not been able to discover.[48] That there ought to be such regulations is suggested in the Edict on Intensifying Nature Protection in Areas of the Far North and Marine Areas adjacent to the Northern Coast of the USSR of 1984,[49] paragraph 1 of which states that in order to protect the natural environment of marine areas adjacent to the northern coast of the USSR 'special rules for the design, construction and operation of . . . installations . . . shall be established'. It may also be noted that under paragraph 7 of the same Edict 'prospecting, geological survey, mining, construction and other work in . . . marine areas adjacent to the northern coast of the USSR must be carried out by means having the least adverse effects on the natural environment'.

Production water

Like drilling-muds, this matter also falls within the remit of the Paris Commission. In 1986 it adopted a non-binding recommendation, setting a maximum limit of 40 g/l of oil in production water.[50] The same comments about the desirability and feasibility of making this standard stricter for the Barents Sea, and about the existence of any Soviet regulations, can be made as with drilling-muds.

Drainage of oil from installations

This matter is dealt with by the International Convention for the Prevention of Pollution from Ships (MARPOL) 1973,[51] Regulation 21 of Annex I of which sets a maximum limit of 100 parts per million of oil to water in oily water draining from installations. Both Norway and the USSR are parties to the Convention. Again, the same comments about the desirability and feasibility of making this standard stricter for the Barents Sea can be made.

Discharge of chemicals used in drilling

This matter is not dealt with by international law. In Norway the matter is dealt with by the 1981 Pollution Law, which applies to the continental shelf. Under this Law any activity causing regular pollution on the continental shelf requires a permit. Thus, oil companies using chemicals on a regular basis in drilling for oil and gas require a permit. Such permits can set limits on the amounts and types of chemicals which may be discharged. In the case of particularly toxic chemicals, a permit can, of course, be refused.

In the USSR, under Resolution No. 564 of 18 July 1969 of the Council of Ministers,[52] there is a general obligation on those carrying out activities on the USSR's continental shelf to prevent pollution of the shelf and its superjacent waters from 'industrial or other commercial refuse, sewage, radioactive substances or industrial waste' (para. 5). In addition, under the Edict on Intensifying Nature Protection in Areas of the Far North and Marine Areas Adjacent to the Northern Coast of the USSR of 1984,[53] the discharge of 'wastes, materials, and articles' in 'marine areas adjacent to the northern coast of the USSR' is prohibited (para. 11). Whether it has been considered necessary to supplement these general obligations with more specific provisions relating to chemicals used in offshore drilling, the authors have been unable to discover.

Discharge of debris, garbage and sewage

These kinds of discharges are dealt with to some degree by international law. Under Regulation 4 of Annex V of the 1973 International Convention for the Prevention of Pollution from Ships[54] the discharge of garbage from offshore oil and gas platforms is prohibited. Both Norway and the USSR have ratified Annex V of the Convention. Annex IV of the Convention deals with the discharge of sewage from ships, but unlike Annex V does not extend to offshore oil and gas platforms. As far as debris is concerned, the London Dumping Convention of 1972[55] prohibits the dumping of certain toxic wastes from ships, and requires a permit to be obtained before other kinds of waste may be dumped. The Convention extends to dumping from offshore oil and gas installations, but does not apply to the disposal of wastes incidental to or derived from the normal operation of installations. Both Norway and the USSR are parties to the Convention.

These international provisions are supplemented at the national level. The Norwegian legislation prohibiting the dumping of debris from the offshore oil industry has already been discussed.[56] The 1981 Pollution Law, mentioned above, is also relevant. In the USSR the 1984 Resolution of the Council of Ministers and 1984 Edict, both discussed in the previous sub-section, are clearly relevant: in particular, the latter prohibits 'the discharge of sewage waters [*sic*] not purified up to the established norms' (para. 11). In addition, the Statute on the Protection and Preservation of the Marine Environment in the Economic Zone of the USSR of 1984[57] provides for the regulation of the discharge of polluting substances from offshore installations in the USSR's economic zone: detailed norms to this end are to be established by the relevant Soviet authorities.

Blow-outs and ruptures of pipelines

So far the forms of pollution we have considered have been deliberate, arising from normal offshore operations. For these forms of pollution regulatory legislation, of the kinds discussed, is clearly the appropriate legal response. The remaining forms of pollution arise from accidents; and their prevention, or at least minimisation, clearly demands a different legal response, where the emphasis is on prescribing various kinds of preventive, rather than regulatory, measures (as well as providing for contingency plans and compensation).

In the case of blow-outs and pipeline ruptures, what is required to prevent such occurrences and the possible resulting pollution is the laying down and rigorous enforcement of strict standards relating to the construction and operation of offshore installations and pipelines. This is a matter for national law: there is no international law on the question. In the case of Norway, without going into any details of its extensive regulations on the topic, it can be said that Norway is generally regarded as having some of the strictest standards in the world, with a reputation for the rigorous enforcement of such standards. In the USSR, on the other hand, while there undoubtedly exist some standards concerning the construction and operation of installations and pipelines,[58] doubts have been expressed over the adequacy of such standards and, in particular, as to how well they are observed and enforced, as mentioned earlier in this chapter.

Collisions between ships and offshore installations

The 1958 Continental Shelf Convention, to which both Norway and the USSR are parties, seeks to reduce the chances of such collisions occurring, first, by prescribing that due notice must be given of the construction of installations and permanent means for giving warning of their presence must be maintained; second, by prohibiting the establishment of installations where interference may be caused to the use of recognised sea-lanes essential to international navigation; and, third, by permitting states to establish 500-metre safety zones around installations from which ships (other than, of course, supply vessels and the like) are normally prohibited. Similar provisions are found in the 1982 UN Convention on the Law of the Sea.

Accidental spillages where oil is being loaded from installations onto ships

The same observations can be made here as in relation to blow-outs and pipeline ruptures, namely that the best way of preventing such accidents is through the establishment and rigorous enforcement of strict construction and operating standards. The same observation can be made, too, about the relative merits of Norwegian and Soviet practice.

Apart from having adequate legislation to regulate operational pollution and minimise the risks of accidental pollution, and ensuring that such legislation is rigorously enforced, there are three or four other matters relating to pollution control which ought to be satisfactorily attended to before commercial production of oil and gas from the Barents Sea begins. First, there needs to be adequate monitoring so as to detect promptly any accidental pollution that occurs and to keep a check on such operational pollution as is permitted. The present Norwegian system of monitoring pollution from its offshore oil and gas operations is regarded as one of the most comprehensive and stringent in the world.[59] What kind of monitoring system (if any) the USSR has for its offshore petroleum industry the authors have been unable to discover, but it is unlikely to be as comprehensive or sophisticated as the Norwegian system. There is obviously room for some collaboration here between Norway and the USSR, and this appears now to be taking place under the framework for co-operation established by their 1988 bilateral Agreement on Environmental Co-operation.[60]

Second, there is a need for both Norway and the USSR to have adequate equipment and manpower close at hand so as to be able to respond promptly and effectively to any accidental pollution that does occur. Again, with its experience in the North Sea and elsewhere, Norway already has the necessary administrative and legislative framework to meet such a goal more effectively, probably, than the USSR does. In addition to individual responses to any accidental pollution, it would be desirable for Norway and the USSR to be able to respond jointly to any such contingency should the circumstances so warrant it, through a previously formulated joint plan of action. Some co-operation to this end is already taking place within the framework of the 1988 Environmental Co-operation Treaty mentioned above, although apparently the absence of an agreed maritime boundary in the Barents Sea is an obstacle to a more comprehensive joint plan of action.[61] An alternative might be the conclusion of a special agreement on the matter, along the lines of the bilateral agreements for Arctic and sub-Arctic waters concluded by Denmark and Canada,[62] Canada and the USA[63] and the USA and the USSR,[64] respectively, in recent years.[65] Again, the absence of an agreed boundary would be likely to be an obstacle, although both the USA–Canada and USA–USSR agreements were concluded at a time when there was no agreed maritime boundary between the states concerned. A third possibility might be for some kind of arrangement[66] to be concluded under the general Agreement on the Protection of the Arctic Environment which is currently being negotiated by the eight Arctic states, assuming of course that such negotiations are successful.

Finally, there is a need for legislation to provide prompt and generous compensation to the victims of any pollution that may occur. Apart from its compensatory purpose, the existence of such legislation may also have the effect of inducing operators of offshore installations to comply as strictly as possible with all the rules and regulations designed to avoid and minimise pollution. In Norway such legislation already exists. Under chapter V of the Petroleum Law of 1985 the licensee of an offshore installation has, subject to some limited exceptions, strict and unlimited liability for pollution damage caused by his installation. As far as the authors have been able to discover, there is no Soviet legislation dealing specifically with compensation for pollution caused by the offshore oil and gas industry.

CONCLUSIONS

Gas has been found in commercial quantities in the Soviet sector of the Barents Sea, and the prospects for making similar finds of gas in the Norwegian sector and of oil in both sectors are quite promising. Exactly how rapid and how extensive commercial production of existing and future finds, and continued exploratory activity, will be is very difficult to predict, depending on such imponderables as political developments in the USSR and the Gulf and the future price of oil and gas. If and when commercial production does begin, there is already a reasonably satisfactory legal regime in place to regulate it. Where there are doubts about the adequacy of the regime, these are in relation to its practical implementation in the Soviet sector and certain deficiencies or possible deficiencies in relation to existing provisions designed to prevent or minimise pollution from offshore oil and gas activities in both sectors of the Barents Sea, in particular the non-participation of the USSR in the Paris Convention and the absence of agreement on contingency plans for large-scale accidental pollution. In addition, uncertainties over the application of the Svalbard Treaty offshore mean that there may be either disputed exploration and exploitation of Svalbard's continental shelf or none at all, or that exploitation may take place under the inadequate provisions of the Mining Code; while the absence of an agreement on a continental shelf boundary both removes a sizeable area from offshore oil and gas activities and hinders the conclusion of effective bilateral contingency plans for combating pollution.

6 Conclusions

INTRODUCTION

It is now time to summarise the main conclusions from the discussion in the preceding chapters. We will also try to undertake a more comprehensive assessment of the different management issues discussed earlier, and present some thoughts about the policy options available.

Security considerations have up to now formed the basis for the kind of resource management measures that could be acceptable. Recently, however, we have experienced the end of the cold war in Europe. Soviet *glasnost* and *perestroika* have also reached the Barents Sea, (e.g. through Mr Gorbachev's Murmansk initiative and increasing Soviet acceptance of Norwegian jurisdiction on Svalbard). The negotiations on arms limitation and disarmament have, however, not shown as much progress for weapons systems at sea as they have on land. Furthermore, geography does not change. Murmansk will continue to be a vital port for the USSR. Overall, however, the general conclusion would be that tension has decreased in the Barents Sea. This means that there is more room for maritime resource management in this area based on management rather than security needs. But security considerations will still continue to play an important part in such management.

FISHERIES

The rich fishery resources of the Barents Sea have been depleted by over-fishing during recent decades. Large natural fluctuations have occurred in the past and are also to be expected in the future. In Chapter 4 we concluded that the prospects for better fishery management have increased with the introduction of the 200-mile regime, although it must be conceded that the size of the North-east Arctic cod stock has continued to decline, even after the establishment of 200-mile zones.

Even so, the 200-mile regime does give a better basis for management than the previous North-east Atlantic Fisheries Commission regime.

A rational future management is above all dependent on the will and ability of Norway and the USSR to pursue a long-term management policy. But success will also depend on such factors as increased scientific knowledge about life in the oceans, the protection of young fish and effective enforcement procedures, as well as appropriate conditions for the management of marine mammals. The 'Grey Zone Agreement' provides satisfactory opportunities for fishery management in the disputed delimitation area. But a solution to the fishery management problems in Svalbard's 200-mile zone has still to be found (see pp. 150–3)

OIL AND GAS ACTIVITY AND ENVIRONMENTAL CONCERNS

Unlike fishing, the exploitation of oil and gas represents a future potential use of the Barents Sea, rather than being a current use. The exploration of the Barents that has so far taken place has shown that gas exists in commercial quantities and indicates that the likelihood of finding oil is quite high. The proper management of any future exploitation of oil and gas requires: first, a regime that provides for orderly exploitation; second, that pollution from exploitation activities should be minimal; and third, that conflicts between oil and gas exploitation and other users of the Barents Sea be kept to a minimum.

The basic elements of a legal regime to provide for the orderly exploitation of oil and gas and to minimise pollution resulting from such exploitation already exist. The major weaknesses and gaps in this regime are doubts about the USSR's capacity effectively to implement its offshore petroleum legislation, which on paper at least appears reasonably adequate; certain deficiencies concerning pollution, in particular the non-participation of the USSR in the Paris Convention and the lack of effective bilateral contingency plans for combating pollution (to which the absence of an agreed continental shelf boundary is a major contributory factor); and, above all, uncertainties over the application of the Svalbard Treaty offshore, which mean that there may be disputed or no exploration and exploitation of Svalbard's continental shelf or that exploitation of Svalbard's continental shelf may take place under the inadequate provisions of the Mining Code.

As far as the final element of a proper management regime is concerned – minimising conflict between oil and gas exploitation and other uses and users of the Barents Sea – the picture is less clear. There are sufficient legal provisions in place, both at the national and international level, to prevent any serious conflicts between shipping and sea-bed resource exploitation, at least as far as civilian shipping is concerned: the position with military shipping is perhaps more problematical. As regards possible conflicts with fishing, this is governed primarily by national law. Norwegian law goes to considerable lengths to minimise and reduce the impact of conflicts between fishing and offshore oil and gas exploitation – although if the latter is to take place a certain level of conflict is inevitable. Soviet law, on the other hand, appears to be completely silent on the issue, and it remains to be seen how the Soviet authorities, through administrative practice, will seek to reconcile their differing interests in fishing and oil and gas exploitation in the Barents Sea.

DELIMITATION OF MARITIME BOUNDARIES

There are no precise rules in international law governing maritime boundary delimitation. It would seem, however, that applying a modified median line as a boundary between the 200-mile economic zones and continental shelves of Norway and the USSR would find support in both the relevant conventions and the case law.

So far, there has been until recently no substantive progress in the negotiations over a boundary between Norway and the USSR. The 'Grey Zone Agreement' provides a rational regime for fisheries management (see pp. 111–14), which means that there is no urgent need for a final delimitation from a fisheries point of view.

The unresolved boundary question prevents oil drilling in the disputed area. This may be considered economically unsatisfactory, but both the environment and the fishing industry may benefit from this stalemate. This situation may, however, cause difficulties if an oilfield is discovered which extends into the disputed area. The unsolved delimitation also creates difficulties in establishing effective contingency arrangements for dealing with large-scale pollution in the Barents Sea.

The boundary issue in the Barents Sea is also a potential source of conflict, like any area with disputed jurisdiction. This is sufficient reason to direct efforts towards finding a solution sooner rather than later. Another aspect of delay in finding a final delimitation is the risk of the 'Grey Zone Agreement' having the effect of bringing the eventual boundary further westwards. It has even been claimed that the fact that the area covered by the 'Grey Zone Agreement' extends west of the sector line may imply that the USSR will not be content with a final delimitation along the sector line, but will claim recognition of certain interests west of the sector line. This could be an argument in favour of Norwegian non-renewal of the agreement and of accepting the sector line as the final delimitation.[1] Our view, however, is that to the extent that the 'Grey Zone Agreement' gives any indication about what the USSR may accept as the final solution, it rather indicates a delimitation between the sector line and the median line; the reason being that a larger part of the area covered by the agreement is to be found east of the sector line than west of this line.[2] As we have stated in Chapter 3, there have also been some other signs recently that the USSR is more willing to compromise. Our general conclusion would therefore be that the most urgent reason for pursuing a solution to the boundary question is the need to eliminate a potential source of conflict between Norway and the USSR.

SVALBARD

We concluded in Chapter 2 that the Svalbard Treaty and the Mining Code apply in Svalbard's territorial sea, on the continental shelf and in the 200-mile zone around Svalbard; but we also argued that, from a political point of view, a solution based on full Norwegian sovereignty was to be preferred. Full Norwegian sovereignty would prevent conflicts over the interpretation of the Svalbard Treaty and it would make it easier to implement effective regulatory measures and enforce them, both on the continental shelf and in the 200-mile zone. Legally, nothing would prevent other parties to the Svalbard Treaty from accepting such Norwegian sovereignty, and consequently abandoning their Treaty rights in the 200-mile zone and on the continental shelf. We will discuss below whether such a solution is politically feasible, as well as the need for an early resolution to the dispute over the application of the Svalbard Treaty and the procedures by which such a resolution might be achieved.

The dispute over the Svalbard Treaty's geographical application has so far prevented the commencement of oil drilling on Svalbard's continental shelf. As with the situation in the grey zone, this is a disadvantage from an economic point of view, but is to the benefit of security interests and fishery and environmental concerns.

We concluded, in Chapter 4, that there is no immediate crisis over fisheries management in the 200-mile zone around Svalbard. The situation is, nevertheless, not satisfactory. First, Soviet vessels do not report their fishing activities in the 200-mile zone. Second, Norway does not arrest fishing vessels violating fishery regulations in the Svalbard zone; such vessels only receive a written warning from the Norwegian

Coast Guard. Third, the disputed legal status of these waters implies that any new conservation measures may meet objections from other states. Since a dispute may be brought before the International Court of Justice, Norway will be careful when implementing such measures. The situation today may thus result in a special status for the USSR; conflicts over the implementation and enforcement of fishery regulations; and the abstention of Norway from the introduction of necessary conservation measures. From a fisheries point of view, there is therefore the need for a solution to the uncertain status of the Svalbard zone soon. We also concluded that, as regards fisheries, Norway can live with the application of the Svalbard Treaty to the 200-mile zone, and consequently that, from a fisheries point of view, Norway should accept such application and commence full enforcement of its regulations, including the arrest of fishing vessels which violate regulations in the zone.

The situation is thus that, although there is no crisis at the time of writing, the need to establish effective fisheries management calls for Norway to accept the Svalbard Treaty's application in the 200-mile zone around Svalbard. Other considerations may, however, call for Norway to pursue its efforts to convince other states about the common interest in a solution based on the non-application of the Svalbard Treaty beyond the territorial sea. Norway must thus balance the need for an early solution based on the Treaty's application in these maritime areas against the possibilities for a future solution based on the Treaty's non-application.

It could be, and has been, argued that it is in other states' interests that the Svalbard Treaty does not apply on the continental shelf and in the 200-mile zone. All states have a general interest in preserving peace and thus in accepting a regime that would reduce the possibilities for conflicts in a strategically sensitive area.[3] This must be a sound argument. On the other hand, the Svalbard regime has been practised on the land areas of Svalbard without creating too many difficulties (although it is true that, because of the inappropriateness of the Mining Code, this regime will be more difficult to practise beyond the territorial sea). The danger to the preservation of peace if the Svalbard Treaty applies beyond the territorial sea must accordingly not be over-estimated. Conversely, the current Norwegian position (that the Treaty does not apply) does not promote low tension in the area, at least as long as this position is disputed.

A further reason why it may be in other states' interests that the Treaty does not apply is that non-application of the Treaty would prevent competition and rivalry over activities on the Svalbard shelf. Such rivalry could result in one party's introduction of generally undesired measures to control another party's expansion.[4] If the Svalbard Treaty did not apply, none of the parties would have any legal basis for claiming a presence on the shelf. On the other hand, the USSR would have no guarantee against a large Western activity, because Western oil companies (or rather their Norwegian subsidiaries) might well be licensed by Norway to operate on the Svalbard shelf, whereas Soviet entities might not be so licensed to any great degree or indeed at all.[5] However, if the Treaty did apply, it does not necessarily follow that there would be competition and rivalry of the kind described. It would seem possible that oil activities on the Svalbard shelf could be delayed until a more orderly system of exploitation had been developed, for example through a revision of the Mining Code.

Norwegian interests are obviously best served with a solution where the Svalbard Treaty does not apply on the continental shelf and in the 200-mile zone: apart from making it easier for Norway to prescribe and enforce regulations, such a solution would also give Norway greater access to the resources of the areas concerned as well as meaning that Norway's powers of taxation would not be limited by article 8 of the Svalbard Treaty. The solution that would best serve other states' interests in the long run is not so obvious, however. The best indication of their opinion would seem to be the lack of development towards their accepting the Norwegian view. As we have concluded above, these states' rejection of or reservation to the Norwegian view has a solid legal basis. Apart from the legal arguments, it seems that other states will give priority to their interest in having equal economic rights to the exploitation of the maritime resources around Svalbard and their right to have a presence in this area. Norwegian prospects for a solution based on the non-application of the Svalbard Treaty therefore seem relatively modest.

Norway may nevertheless still pursue its efforts to convince other states of their interest in the 'Norwegian' solution. But we would argue that such a policy should not continue if the cost would be uncontrolled fishing in the 200-mile zone around Svalbard. Fisheries developments in the 200-mile zone must therefore be monitored on a continuous basis and Norwegian Svalbard policy should be adjusted if need be.

Different procedures to obtain its preferred solution may be followed by Norway in the time ahead. One possibility would be to have the legal dispute settled by the International Court of Justice. This would formally only be binding for the parties involved in the case, but would give political legitimacy to the solution found by the Court. As we have concluded in Chapter 2, we believe that the Court would accept that Norway had sovereign rights on the continental shelf and in the 200-mile zone, but decide that those rights were subject to other states' rights under the Svalbard Treaty. Accordingly, Norway would not be interested in bringing a case before the Court as a means of resolving the dispute.

Theoretically a new Svalbard Conference could be convened to adjust the Svalbard Treaty to the management needs of today. The outcome of such a conference would, however, be unpredictable and it would probably be more likely to produce uncertainty than an improved Treaty. We would therefore not recommend this alternative. It would thus seem that an informal political process is to be preferred if Norway is to stand a chance of convincing other states of the virtues of its preferred solution.

Norway has traditionally favoured treating the different disputes in the North as separate issues,[6] whereas the USSR may prefer a package deal.[7] It has, however, been argued that Norway may also benefit from linking the delimitation and Svalbard questions and that a package could be that the USSR accepted the Norwegian view on the Svalbard Treaty's non-application beyond the territorial sea, while Norway recognised a dividing line 'closely approximating the sector line'.[8] The difficulty with such a package, however, is that the USSR cannot commit other states' views on the Svalbard Treaty's geographical application. This package would thus not solve the problems Norway is having with other states objecting or making reservations to the Norwegian view. A more realistic package deal with the USSR would be that Norway accepted that the Svalbard Treaty applied beyond the territorial sea in return for Soviet agreement on a boundary line close to the median line. However, it is not likely that Norway would yield on the Svalbard question in order to gain in the boundary dispute. The benefit of claiming that each case should be solved separately is that this approach is more likely to promote solutions based on international law; as is well known, there is no international law on package deals. We would therefore recommend that this approach be continued.

Notes

The following abbreviations have been used throughout:

ICJ Rep.:	International Court of Justice Reports (year of publication in square brackets)
PCIJ:	Permanent Court of International Justice
RIAA:	Reports of International Arbitral Awards

1 INTRODUCTION

1 Throughout this book we use the term 'Svalbard' to refer to this territory, which is the modern Norwegian name, now increasingly used by non-Norwegians; rather than Spitsbergen, which is the older English name and the name used in the Treaty. Spitsbergen is now used to refer to the largest island in the archipelago, which was formerly called West Spitsbergen. Likewise the Treaty concerning the Archipelago of Spitsbergen is referred to informally in this book as the Svalbard Treaty.

2 K. Traavik and W. Østreng, 'Security and Ocean Law: Norway and the Soviet Union in the Barents Sea', *Ocean Development and International Law*, IV (1977), pp. 343–67 at 346.

3 Ibid.

4 E. Bergsager, 'Barentshavet: Ressurser og Utvinningsmuligheter', *Internasjonal Politikk*, 2–3 (1988), pp. 7–34 at 10.

5 W.E. Butler, *Northeast Arctic Passage*, Alphen aan den Rijn: Sijthoff and Noordhoff, 1978, pp. 9–10; H.O. Bergesen, A. Moe and W. Østreng, *Soviet Oil and Security Interests in the Barents Sea*, London: Frances Pinter, 1987, pp. 58–9.

6 D.G. Groves and L.M. Hunt, *Ocean World Encylopaedia*, New York: McGraw-Hill, 1980, p. 30.

7 *Encyclopaedia Britannica*, 15th edn, Micropedia, I, p. 894.

8 Groves and Hunt, op. cit., p. 29.

9 And its motive for doing so is less economic than the political one of maintaining a presence on Svalbard to keep an eye on what is going on there and to ensure that nothing takes place which might adversely affect its security interests.

10 The information in this and the previous paragraph is largely taken from the

Encylopaedia Britannica, 15th edn, Micropaedia, IV, p. 943 and VIII, pp. 809–10.

11 Groves and Hunt, op. cit., p. 29.

12 International Council for the Exploration of the Sea (hereafter ICES), *Reports of the ICES Advisory Committee on Fishery Management, 1989*, ICES Co-operative Research Report 168, Charlottenlund, Denmark, 1990, pt II, p. 85.

13 The account that follows is very largely based on various publications of ICES, especially *Survey of Fish Resources in the North-east Atlantic*, ICES Co-operative Research Report 37, 1974; the *Reports of the ICES Advisory Committee on Fishery Management, 1989*, and *Bulletin Statistique des Pêches Maritimes*, LXXII, 1990.

14 All the figures just given relate to all catches for this stock, i.e. including catches outside the Barents Sea. It is difficult to give figures for the Barents Sea only, because many stocks (as indicated) migrate outside the Barents and also because the ICES Statistical Areas to which catches refer do not correspond exactly with the boundaries of the Barents. In the account which follows the authors have nevertheless, unless otherwise indicated, tried to calculate catches for the Barents Sea only.

15 For a useful summary of such knowledge, see M. Rider, 'Play Potentials of the Barents Shelf', *Oil and Gas Journal*, LXXXVI/4 (1988), pp. 87–90.

16 For a fuller account of Soviet offshore oil and gas exploratory activities, see Bergesen, Moe and Østreng, op. cit., ch. 2.

17 *Offshore*, XLIX/10 (1989), p. 7 and *Offshore*, XLIX/11 (1989), p. 7.

18 Strategic issues in Northern waters have generated a vast literature. Among more recent and important contributions see: C. Archer and D. Scrivener (eds), *Northern Waters: Security and Resource Issues*, London: Croom Helm, 1986; C. Archer (ed.), *The Soviet Union and Northern Waters*, London: Routledge, 1988; Bergesen, Moe and Østreng, op. cit., ch. 3; G. Lindsey, *Strategic Stability in the Arctic*, London: Brassey, 1989; K. Mottolä (ed.), *The Arctic Challenge*, Boulder, Colo.: Westview Press, 1988; W. Østreng, *The Soviet Union in Arctic Waters*, Hawaii, Honolulu: The Law of the Sea Institute (Occasional Paper 36), 1987.

19 Apart from the military activities referred to specifically in the previous section, it should also be mentioned that the USSR has from time to time used areas of the Barents Sea for weapons testing.

20 Such lines were first drawn in 1869. The current legislation laying down these lines is the Royal Decree of 12 July 1935. English translation in UN Legislative Series B/6, p. 35.

21 Royal Decree of 25 September 1970. English translation in UN Legislative Series B/15, p. 21.

22 Ibid.

23 Decree of the Council of Ministers of 15 January 1985. English translation in W.E. Butler, *The USSR, Eastern Europe and the Development of the Law of the Sea*, Dobbs Ferry, NY: Oceana, 1987, booklet C3.

24 Ibid.

25 The current Soviet legislation is the Regulations of 5 August 1960 for the Protection of the State Frontier of the USSR, as amended in 1971, art. 3. English translation in UN Legislative Series B/16, p. 34.

26 Royal Decree of 22 February 1812. English translation in UN Legislative

Series B/6, p. 35. That the four-mile limit applies to Svalbard was confirmed by the Royal Decree of 25 September 1970, see above.

27 Royal Decree of 31 May 1963. English translation in UN Legislative Series B/15, p. 393. Although the point is not free from doubt, the Decree appears to apply to Svalbard: see R.R. Churchill and P. Ørebech, 'Om norsk lovgivnings anvendelse på Svalbard', *Lov og Rett*, (1985), pp. 164–77 at 172–3.

28 Decree of the Presidium of the Supreme Soviet of the USSR of 6 February 1968. English translation in UN Legislative Series B/15, p. 441.

29 Petroleum Law of 22 March 1985, art. 4. This Law appears to apply to Svalbard: Churchill and Ørebech, op. cit., p. 174.

30 As was confirmed by the International Court of Justice in 1985 in the *Continental Shelf (Libya/Malta)* case, [1985] ICJ Rep., pp. 13 at 33.

31 Law No. 91 of 17 December 1976. English translation in UN Legislative Series B/19, p. 241.

32 Royal Decree of 3 June 1977. *Norsk Lovtidend: 1977*, pt I, p. 508.

33 Decree of 10 December 1976. English translation in UN Legislative Series B/19, p. 253. The Decree was applied to the Barents Sea in May 1977.

34 Decree of 28 February 1984. English translation in the UN's *Law of the Sea Bulletin*, 4 (1984), p. 32.

2 APPLICATION OF THE SVALBARD TREATY TO MARITIME ARFAS

1 T. Mathisen, *Svalbard i Internasjonal Politikk 1871–1925*, Oslo: H. Aschehoug & Co., 1951, p. 161.

2 *Svalbard*, Norwegian Government Paper Stortingsmelding 40, 1985–6, pp. 39–40.

3 E. Bergsager, 'Barentshavet: Ressurser og Utvinningsmuligheter', *Internasjonal Politikk*, 2–3 (1988), p. 31.

4 *Åpning av Barentshavet Syd for letevirksomhet*, Norwegian Government Paper Stortingsmelding 40, 1988–9, p. 15.

5 *UD-informasjon*, 24, 1987, p. 27.

6 *UD-informasjon*, 22, 1987, p. 41.

7 See Stortingsmelding 40, 1988–9; Innstillinger til Stortinget S. 216 1988–9; Stortingstidende, pp. 4197–221 (7 June 1989).

8 *Om godkjennelse av traktaten om Spitsbergen (Svalbard)* Norwegian Government Paper Stortingsproposisjoner 36 (1924), p. 7.

9 Ibid., p. 9.

10 Mathisen, op. cit., p. 117.

11 League of Nations Treaty Series, II, p. 8. The treaty entered into force in 1925 and has been ratified or acceded to by about forty states, including the United States, the USSR, Japan, and most European states.

12 For a full account of the background to and negotiation of the Svalbard Treaty, see W. Østreng, *Økonomi og Politisk Suverenitet, Interessespillet om Svalbards Politiske Status*, Oslo: Universitetsforlaget, 1974.

13 I. Brownlie, *Principles of Public International Law*, Oxford: Oxford University Press, 3rd edn, 1979, p. 624; Lord McNair, *The Law of Treaties*, Oxford: Clarendon Press, 1961, p. 366; and H. Lauterpacht, 'Restrictive Interpretation and the Principle of Effectiveness in the Interpretation of

Treaties', *British Yearbook of International Law*, XXVI (1949), pp. 48–85 at 62.

14 Stortingsmelding 39, 1974–5, p. 7. This is supported by Professor Fleischer, Legal Adviser to the Norwegian Ministry of Foreign Affairs, see C.A. Fleischer, *Petroleumsrett*, Oslo: Universitetsforlaget, 1983, pp. 194–5 and 'Le régime d'exploitation du Spitsberg (Svalbard)', *Annuaire français de droit international*, XXIV (1978), pp. 275–300 at 279–80.

15 *Grisbådarna Case*, RIAA, XI (1909), p. 147; *Island of Palmas Case*, RIAA, II (1928), p. 829; *Right of Passage over Indian Territory*, [1960] ICJ Rep., pp. 6 ff.; *Western Sahara Case*, [1975] ICJ Rep., pp. 12 ff.; *Namibia Case*, [1971] ICJ Rep., pp. 16 ff.; and *Aegean Sea Continental Shelf Case*, [1978] ICJ Rep., pp. 3 ff..

16 The principle of contemporaneous interpretation is discussed *inter alia* in McNair, op. cit., p. 467, D.P. O'Connell, *International Law*, London: Stevens & Sons, 2nd edn, 1970, pp. 257–8, Brownlie, op. cit., p. 626 and T.O. Elias, 'The Doctrine of Intertemporal Law', *American Journal of International Law*, LXXIV (1980), pp. 285–307.

17 *North Atlantic Coast Fisheries Case*, 1 Scott, Hague Court Reports, 141 (1916); *Kronprinz Gustaf Adolf Case*, RIAA, II (1932), p. 1239 ff. at 1254; *Radio Corporation of America* v. *The National Government of the Republic of China*, RIAA, III (1935), pp. 1621, 1935–7.

18 *The S.S. 'Wimbledon'*, PCIJ, Ser. A, 1 (1923); *Mosul Case*, PCIJ, Ser. B, 12 (1925), p. 25; *Territorial Jurisdiction of the International Commission of the River Oder*, PCIJ, Ser. A, 23 (1925); *Free Zones of Upper Savoy and the District of Gex (second phase)*, PCIJ, Ser. A, 24 (1932); *Free Zones of Upper Savoy and the District of Gex*, PCIJ, Ser. A/B, 46 (1932); and *Interpretation of the Statute of the Memel Territory,* PCIJ, Ser. A/B, 49 (1932).

19 PCIJ, Ser. A, 1 (1923), p. 24–5. See also the *Mosul* case, PCIJ, Ser. B, 12 (1925), p. 25.

20 PCIJ, Ser. A, 1 (1923), p. 23.

21 The *Memel* case, PCIJ, Ser. A/B, 49 (1932), p. 333.

22 Ibid., p. 335.

23 *Oder* case, PCIJ, Ser. A, 23 (1925), p. 26.

24 M.S. McDougal et al., *Interpretation of Agreements and World Public Order*, New Haven and London: Yale University Press, 1967, p. 156.

25 Lauterpacht, op. cit., p. 67.

26 Cf. *Reparations for Injuries Suffered in the Service of the United Nations*, [1949] ICJ Rep., pp. 174 ff. and *Certain Expenses of the United Nations Case*, [1962] ICJ Rep., pp. 151 ff.. McDougal *et al.*, op. cit., p. 161, also refers to the following judgments of the Permanent Court: *Greco-Bulgarian 'Communities' Case*, PCIJ, Ser. B, 17 (1930), *Certain Questions Relating to Settlers of German Origin in the Territory Ceded by Germany to Poland*, PCIJ, Ser. B, 6 (1923), *Aquisition of Polish Nationality*, PCIJ, Ser. B, 7 (1923).

27 Cf. McDougal *et al.*, p. 185:

> The principle of effectiveness and restrictive interpretation are therefore two different sides of the same fundamental policy – that of securing and preserving the integrity of agreement by reference to the major objectives of the parties. The one principle is a positive formulation that ambiguities and contradictions in specific provisions should be complemented and

remedied in a way to promote the essential purposes of the parties; the other principle is a negative formulation that implementing inferences should be made to stop short of defeating these purposes, and of imposing unassumed burdens either of general or detailed obligation.

28 *Commission du Spitsberg*, Recueil des actes de la Conférence, pt VII, Conférence de la paix, 1919–20, Paris 1924, p. 90.
29 *Collection of Laws &c. 1921–25 Relating to the Commerce and Shipping of the Kingdom of Norway &c*, The Norwegian Ministry of Foreign Affairs, Oslo: Grøndahl & Søn, 1928, pp. 853–85.
30 It is claimed that the Mining Code is binding in international law by F. Castberg, *Utredninger om folkerettslige spørsmål 1922–41*, Oslo: The Norwegian Ministry of Foreign Affairs, 1950, p. 380 and J. Andenæs, *Suverenitet og Eiendomsrett på Svalbard*, Oslo: Universitetsforlaget, 1984, p. 9. C.A. Fleischer, op. cit., pp. 179, 200, 286 respectively, takes the opposite view.
31 Cf. O'Connell, op. cit., p. 205.
32 Arts 2(1), 2(2), 3(1), 3(2).
33 Arts 3(3), 4, 5, 7(1), 8(1), 8(2), 9, 10 (2).
34 See also J.P.A. Bernhardt, 'Spitzbergen: Jurisdictional Friction over Unexploited Oil Reserves', *California Western International Law Journal*, IV (1973), pp. 65–120.
35 See ibid., p. 66.
36 *Commission du Spitsberg*, pp. 58, 67.
37 See n. 33 above.
38 United Nations Treaty Series, vol. DXVI, p. 206.
39 RIAA, XI (1909), p. 147 ff. at 159.
40 *Commission du Spitsberg*, p. 44.
41 W. Østreng, *Det Politiske Svalbard*, Oslo: Gyldendal Norsk Forlag, 1975, p. 50.
42 Cf. C.A. Fleischer, 'Oil and Svalbard', *Nordisk Tidsskrift for International Ret*, XLV/1 (1976), pp. 7–13 at 10.
43 J. No. ID 742/67 LS/BD.
44 Fleischer, *Petroleumsrett*, p. 212.
45 The only reference to the continental shelf in the *travaux préparatoires* is in Norway's letter to the Versailles Conference of 10 April 1919, where it is stated that the Bear Island is situated on the same continental shelf as Spitsbergen ('Enfin, au Spitsberg se rattache l'île aux Ours ou Beeren-Eiland, située sur le même plateau sous-marin que cet archipel', (*Commission du Spitsberg*, p. 116)). This point was, however, made by Norway to ensure that Bear Island was included in the Svalbard Treaty. This statement was made long before the continental shelf developed as a legal concept and it has no effect on the Treaty's application on the continental shelf.
46 Art. 10 of the 1958 Convention on the Territorial Sea and Contiguous Zone. Present-day customary international law is generally taken as being the same as the Convention. See R.R. Churchill and A.V. Lowe, *The Law of the Sea*, Manchester: Manchester University Press, 2nd edn, 1988, p. 41.
47 Art. 1 of the 1958 Geneva Convention on the Continental Shelf. In the *North Sea Continental Shelf Cases*, [1969] ICJ Rep., pp. 3 ff., the ICJ stated that art. 1 also represents customary international law.
48 This is generally taken to be the position in customary international law. See Churchill and Lowe, p. 41.

49 Ibid., p. 135.

50 H. Vindenes, 'The Changing Legal Regimes: A Norwegian Perspective', in J.J. Holst, K. Hunt and A.C. Sjaastad, *Deterrence and Defense in The North*, Oslo: Norwegian University Press, 1985, p. 209.

51 Stortingsmelding 40, 1985–6, p. 9.

52 Stortingsmelding 30, 1973–4, p. 67.

53 C.A. Fleischer, 'The Northern Waters and the New Maritime Zones', *German Yearbook of International Law*, XXII (1979), pp. 100–18 at 109.

54 Cf. also Bernhardt, op. cit., p. 96.

55 See Churchill and Lowe, pp. 53–8; C.J. Colombos, *The International Law of the Sea*, London: Longmans, 6th edn, 1967, pp. 88–91.

56 H. Vindenes, op. cit., p. 211 claims: 'in the Norwegian view it is futile to try to find the solution through speculation about what the parties would have done if they had known in 1920 of the future developments of the Law of the Sea'.

57 *Commission du Spitsberg*, pp. 17–19.

58 C.A. Fleischer, *Petroleumsrett*, p. 215.

59 Ibid., p. 223.

60 Ibid., pp. 223–4.

61 Such a boundary is necessary because the EEC's Common Fishery Policy does not apply in its entirety to the Isle of Man's territorial sea and fishing zone.

62 Bernhardt, op. cit., p. 93, seems to claim that Norway is obliged to enter into 'meaningful and sincere negotiation' with the other parties to the Treaty on delimitation.

63 Royal Decree of 3 June 1977, art. 1. *Norsk Lovtidend: 1977*, pt I, p. 508. The nearest part of Svalbard to the Norwegian mainland is Bear Island, some 220 nautical miles north of the North Cape.

64 C.A. Fleischer, 'Oil and Svalbard', p. 10.

65 *Anglo-Norwegian Fisheries Case*, [1951] ICJ Rep., pp. 116 ff. at 133.

66 *North Sea Continental Shelf Cases*, [1969] ICJ Rep., pp. 3 ff. at 23.

67 Ibid., p. 52.

68 *Aegean Sea Continental Shelf Case*, (Jurisdiction), [1978] ICJ Rep., pp.3 ff. at 36.

69 *Libya/Malta Continental Shelf Case*, [1985] ICJ Rep., pp. 13 ff. at 41.

70 *International Law Reports*, XVIII (1951), p. 144.

71 Ibid., p. 161.

72 [1978] ICJ Rep., pp. 3 ff. at 36.

73 *International Law Reports*, XVIII (1951), p. 152.

74 Ibid., p. 149.

75 Ibid., p. 150.

76 [1978] ICJ Rep., p. 32.

77 This conclusion has also been reached by T. Scovazzi, 'Antichi e recenti problemi del regime giuridico delle Spitzbergen', *Studi in onore di Cesare Grassetti*, III (1980). Similarly E. Morelli in his book *Il regime giuridico dello Svalbard e il Nuovo Diritto del Mare*, Milano: Dott. A. Giuffrè, Editore, 1988.

On the other hand, J.P.A. Bernhardt, op. cit., p. 115, concludes on the basis of the Abu Dhabi and the Qatar cases:

> The *rationes decidendi* of Lords Asquith and Radcliffe leave us with but one result: that under article 3 of the Treaty the United States and the

> other High Contracting Parties to the Treaty on the Status of Spitzbergen of 9 February 1920 have never enjoyed rights to oil and natural gas deposits in the continental shelf of Spitzbergen.

He adds however:

> The conclusions reached in this article are not the only ones which can be drawn from the matrix of fact and circumstance which Secretary of State Lansing called 'A Unique International Problem' as early as 1917.

78 Scrivener claims: 'References to Western plans to install acoustic and electronic listening devices covertly in the Svalbard archipelago probably reflect a genuine concern in addition to serving a useful, if little-noticed, propaganda function. The Soviets have themselves in all probability installed seabed acoustic devices throughout the Barents Sea, including the waters of the Svalbard archipelago': D. Scrivener, 'The Soviet Union and Northern Waters', in C. Archer and D. Scrivener (ed.), *Northern Waters: Security and Resource Issues*, London: Croom Helm, 1986, pp. 208–33 at 211.

79 Østreng argues, not very convincingly in our view, that the application of art. 9 to zones beyond the territorial sea would have a significantly limiting effect on Soviet naval activities in the Barents Sea. See W. Østreng, 'The Continental Shelf Issues in the "Eastern" Arctic Ocean: Implications of UNCLOS III, with Special Reference to the Informal Composite Negotiating Text', in J.K. Gamble (ed.), *Law of the Sea: Neglected Issues*, Hawaii: Law of the Sea Institute, 1979, pp. 165–82 at 174.

80 Stortingsmelding 40, 1985–6, p. 9.

81 Petroleum Law of 22 March 1985 No. 11, art. 1(4). Cf. Odelstingsproposisjon 72, 1982–3, p. 37.

82 Stortingsmelding 40, 1985–6, p. 9 and *UD-informasjon*, 22 (1987), p. 40.

83 On behalf of the British Government, Baroness Young stated in the House of Lords: 'In our view Svalbard has its own continental shelf, to which the regime of the Treaty of Paris applies. The extent of this shelf has not been determined'. (House of Lords Debates, CDLXXVII, col. 1022, 2 July 1986).

84 *UD-informasjon*, 24 (1987), p. 27.

85 Stortingsmelding 40, 1985–6, p. 9.

3 BOUNDARY DELIMITATION IN THE BARENTS SEA

1 Agreement concerning the Sea Frontier between Norway and the USSR in the Varangerfjord, Oslo: 15 February 1957. In force 24 April 1957. United Nations Treaty Series, CCCXII, p. 289. The text of the Agreement is also reproduced and analysed in The Geographer, US Department of State, International Boundary Study Series A, 17 *Limits in the Seas*.

2 United Nations Treaty Series, CDXCIX, p. 311.

3 *International Legal Materials*, XXI (1982), p. 1245.

4 Art. 311 of the 1982 Convention.

5 [1969] ICJ Rep., pp. 3 ff.

6 [1982] ICJ Rep., pp. 18 ff.

7 [1984] ICJ Rep., pp. 246 ff.

8 [1985] ICJ Rep., pp. 13 ff.

9 *International Legal Materials*, XVIII (1979), p. 398.

10 *International Legal Materials*, XXV (1986), p. 252.

11 Other maritime boundary arbitrations include the unpublished award concerning the border between the Emirates of Dubai and Sharjah (1981) and the Guinea Bissau/Senegal Maritime Boundary Arbitration Award (1989), *Revue Générale de Droit International Public*, XCIV (1990), p. 204.

12 P. Weil, *The Law of Maritime Delimitation: Reflections*, Cambridge: Grotius Publications, 1989, esp. pp. 3–15 and 149–56.

13 For fuller accounts than that given here, see (apart from the magisterial work by Weil just referred to) among others P.B. Beazley, 'Maritime Boundaries: A Geographical and Technical Perspective', *Law of the Sea Institute Proceedings*, XIX (1985), pp. 319–39; P. Bravender-Coyle, 'The Emerging Legal Principles and Equitable Criteria Governing the Delimitation of Maritime Boundaries between States', *Ocean Development and International Law*, XIX (1988), pp. 171–227; J.I. Charney, 'The Delimitation of Ocean Boundaries', *Ocean Development and International Law*, XVIII (1987), pp. 497–531; M.D. Evans, *Relevant Circumstances and Maritime Delimitation*, Oxford: Clarendon Press, 1989; B. Kwiatkowska, 'Equitable Maritime Boundary: A Legal Perspective', *International Journal of Coastal and Estuarine Law*, III (1988), pp. 287–304; and L.A. Willis, 'From Precedent to Precedent: The Triumph of Pragmatism in the Law of Maritime Boundaries', *Canadian Yearbook of International Law*, XXIV (1986), pp. 3–60.

14 [1969] ICJ Rep., p. 54.

15 Cf. Kwiatkowska, op. cit., pp. 303–4: 'equitable principles do not lay down the obligations but simply clarify the guidelines for achieving an equitable result in delimitation'.

16 R.R. Churchill and A.V. Lowe, *The Law of the Sea*, Manchester: Manchester University Press, 2nd edn, 1988, p. 127.

17 [1985] ICJ Rep., pp. 13 ff. at 33.

18 Convention on the Continental Shelf, art. 2(3) and UN Convention on the Law of the Sea, art. 77(2). In the *North Sea Continental Shelf* cases the International Court of Justice held this also to be the position in customary international law: [1969] ICJ Rep., pp. 3 ff. at 23.

19 Royal Decree of 31 May 1963. English translation in UN Legislative Series B/15, p. 393. In 1985 Norway amended its claim as far as the outer limit of its continental shelf was concerned from the exploitability criterion of the 1963 Royal Decree to the formula of the 1982 UN Convention on the Law of the Sea, i.e. 200 miles or the edge of the continental margin: see Petroleum Law of 22 March 1985, art. 4.

20 Decree of the Presidium of the Supreme Soviet of the USSR of 6 February 1968. English translation in UN Legislative Series B/15, p. 441.

21 P. Tresselt, 'Norsk–sovjetiske forhandlinger om avgrensning av kontinentalsokler og økonomiske soner', *Internasjonal Politikk* (1988), pp. 75–93 at 78–9. This article is of especial importance, as the author was formerly head of the Department of Legal Affairs in the Norwegian Foreign Office and has been involved in the boundary negotiations.

22 The Norwegian economic zone was established by Law No. 91 of 17 December 1976, English translation in UN Legislative Series B/19, p. 241. The Soviet Union originally established a 200-mile fishing zone (see Decree of 10 December 1976, English translation in ibid., p. 253), extended to the

Barents Sea in May 1977, but replaced this in 1984 with a 200-mile economic zone (see Decree of 28 February 1984, English translation in *Law of the Sea Bulletin* 4, (1984), p. 32). The Norwegian legislation provides that the Norwegian economic zone shall not extend beyond the median line in relation to other states; the Soviet legislation that the delimitation of the Soviet economic zone in relation to neighbouring states 'shall be effected . . . by agreement on the basis of international law, in order to achieve an equitable solution', i.e. the 1982 UN Convention on the Law of the Sea formula.

23 The Norwegian Government took the view that the boundaries ought to coincide: see Norwegian Government Papers Stortingsproposisjon 74, 1976–7, p. 3 and Norwegian Government Papers Stortingsproposisjon 70, 1977–8, p. 1.

24 The official title of the Agreement is: Agreement on an Interim Practical Arrangement for Fishing in an Adjoining Area in the Barents Sea. An English translation of this Agreement does not appear to have been published: the Norwegian text can be found in [1978] *Overenskomster med fremmede Stater*, 436. Quotations from the Agreement have been translated by the authors.

25 Norwegian Government Papers Stortingsproposisjon, 70, 1977–8, p. 3.

26 In spite of this provision, some Soviet writers have apparently argued that the Agreement undermines Norway's claim that the boundary should be an equidistant line: see D. Scrivener, *Oil, Fish and Strategy: The USSR, Svalbard and the Barents Sea*, Aberdeen, Aberdeen Studies in Defence Economics No. 27, 1987, p. 18. So, too, have some Norwegian writers: see, for example, J.K. Skogan, 'Gråsoneordningen og fremtiden', *Internasjonal Politikk* (1978), pp. 459–69 at 461. From a strictly legal view this seems difficult to accept – the 'without prejudice' clause must be read in good faith at face value. If it were not, one could as well argue that the Agreement undermines the USSR's claim that the sector line should be the boundary as that it undermines Norway's claim. On the other hand, from a political standpoint it can be argued that by accepting that the Agreement applies to a much larger area of undisputed Norwegian economic zone than undisputed Soviet zone, Norway appears to be more willing to make concessions on boundary questions in future negotiations.

27 Norwegian Government Papers Stortingsproposisjon, 70, 1977–8, p. 3.

28 P. Tresselt, op. cit., p. 84.

29 On Mr Gorbachev's Murmansk speech, see C. Archer, 'Russia's Arctic Dimension', *The World Today*, XLIV/3 (1988), pp. 47–50; D. Scrivener, *Gorbachev's Murmansk Speech: The Soviet Initiative and Western Response*, Oslo: The Norwegian Atlantic Committee (DNAK), 1989.

30 Tresselt, op. cit., pp. 88–90.

31 P.A. Madsen, 'Gorbachev, Hostile Images and the Northern Areas', *International Challenges*, 1989, pp. 15–22 at 20.

32 See Tresselt, op. cit., esp. pp. 85–8.

33 Ibid., p. 85.

34 *UD-informasjon*, 7 (1978); Tresselt, op. cit., p. 85.

35 [1984] ICJ Rep., pp. 301–3.

36 Tresselt, op. cit., pp. 79, 87.

37 Decree of 15 January 1985. English translation in W.E. Butler, *The USSR, Eastern Europe and the Development of the Law of the Sea*, Dobbs Ferry, NY: Oceana, 1987, booklet C3.

38 See *North Sea Continental Shelf* cases, [1969] ICJ Rep., p. 52; *Anglo-French Continental Shelf* case, *International Legal Materials*, XVIII (1979), p. 398, paras 98–101; *Continental Shelf (Tunisia/Libya)* case, [1982] ICJ Rep., p. 75; *Gulf of Maine* case, [1984] ICJ Rep., pp. 322–3, 334–7; *Guinea/Guinea Bissau* arbitration, *International Legal Materials*, paras 118–20; and *Continental Shelf (Libya/Malta)* case, [1985] ICJ Rep., pp. 43–6, 49–50.

39 Tresselt, op. cit., p. 79.

40 Note, however, Weil's formula for determining the relevant coastlines. 'Since delimitation presupposes an over-lapping of projections, the relevant segments [of coastline] for delimitation purposes are those whose projections intersect projections from the other coast': Weil, op. cit., p. 72. Weil in fact goes further than most writers in arguing that proportionality should not be a relevant criterion in delimitation: see ibid. pp. 235–44.

41 See *North Sea Continental Shelf* cases, [1969] ICJ Rep., pp. 31–2; *Continental Shelf (Tunisia/Libya)* case, [1982] ICJ Rep., pp. 61–2; *Gulf of Maine* case, [1984] ICJ Rep., pp. 298–9, 313, 328; *Guinea/Guinea Bissau* arbitration, *International Legal Materials*, XXV (1986), p. 252, paras 103–7.

42 Tresselt, op. cit., pp. 79–80.

43 [1985] ICJ Rep., p. 35.

44 *Gulf of Maine* case, [1984] ICJ Rep., p. 327.

45 See *Libya/Malta* case, [1985] ICJ Rep., pp. 40–1, *Guinea/Guinea Bissau* arbitration, *International Legal Materials*, XXV (1986), p. 252, paras 118–9.

46 In the *Gulf of Maine* case the International Court dismissed as irrelevant the USA's arguments based on the ecology of the Gulf and Canada's invocation of its economic interests: [1984] ICJ Rep., pp. 316–9, 341–3.

47 *International Legal Materials*, XXV (1986), p. 252, para. 124.

48 [1985] ICJ Rep., p. 42.

49 Tresselt, op. cit., p. 80.

50 Decree of the Presidium of the Central Executive Committee of the USSR, 16 April 1926. *SZ SSSR* (1926), no. 32 item 203. English translation in M.O. Hudson, *Cases on International Law*, 1951, p. 220.

51 The reasons for the decree were fears that the USA and Canada might try and claim Wrangel and Herald islands, and possible fears that Norway might claim Franz Josef Land: all these islands lie within the sector referred to in the decree. See Tresselt, op. cit., p. 80.

52 Tresselt, op. cit., pp. 80–1.

53 *Continental Shelf (Tunisia/Libya)* case, [1982] ICJ Rep., pp. 83–6; *Gulf of Maine* case, [1984] ICJ Rep., pp. 303–12; and *Guinea/Guinea Bissau* arbitration, *International Legal Materials* XXV (1986), p. 252, paras 105–6.

54 D. Pharand, *Canada's Arctic Waters in International Law*, Cambridge: Cambridge University Press, 1988, chaps.1–4.

55 *Consolidated Treaty Series*, LXXV, p. 96.

56 *Consolidated Treaty Series*, CXXXIV, p. 332. In their Agreement on the Maritime Boundary (*International Legal Materials*, XIX (1990), p. 941) the USSR and USA have accepted the line of the 1867 Convention as the maritime boundary.

57 Scrivener, *Oil, Fish and Strategy*, p. 12

58 Ibid.

59 [1984] ICJ Rep., p. 320. A similar view appears to have been taken by the Court in the *Tunisia/Libya* case. Cf. also Weil, op. cit., pp. 59–60, 274–5.

60 For a further discussion of this solution, with cartographical illustrations, see A.H. Klepsvik, 'Norges kontinentalsokkel og eksklusive økonomiske sone', *Internasjonal Politikk* (1979), pp. 343–92 at 369–76. One of the problems with using the perpendicular as the boundary, as both Klepsvik and Weil point out, is determining the general direction of the coast. Any slight difference in determining such direction will, of course, markedly affect the course of the boundary, especially at its furthest point from the coast.

61 Tresselt, op. cit., p. 79.

62 See the *North Sea Continental Shelf* cases, [1969] ICJ Rep., pp. 36–7; *Anglo-French Continental Shelf* arbitration, *International Legal Materials*, XVIII (1979), p. 398 para. 239; *Continental Shelf (Tunisia/Libya)* case, [1982] ICJ Rep., p. 88; *Gulf of Maine* case, [1984] ICJ Rep., p. 334; *Continental Shelf (Libya/Malta)* case, [1985] ICJ Rep., p. 47.

63 For writers opposed to the use of ice-shelves, see, for example, F.M. Auburn, *Antarctic Law and Politics*, London: Hurst, 1982, p. 36 and D.P. O'Connell, *The International Law of the Sea*, Oxford: Clarendon Press, 1982, I, pp. 197–8. For the opposite view, see D. Pharand, *The Law of the Sea of the Arctic*, Ottawa: University of Ottawa Press, 1973, pp. 182–8 and the Soviet writer W.L. Lakhtine, 'Rights over the Arctic', *American Journal of International Law*, XXIV (1930), pp. 703–17 at 712.

64 A.O. Adede, *The System for Settlement of Disputes under the United Nations Convention on the Law of the Sea*, Dordrecht: Martinus Nijhoff, 1987, pp. 179–82, and S.P. Jagota, *Maritime Boundary*, Dordrecht: Martinus Nijhoff, 1985, p. 237.

65 On the other hand, prolonged continuation of the boundary dispute may lead to increased tension in the area, which is not in the Soviet Union's security interests: see W. Østreng, 'Delimitation Arrangements in Arctic Seas', *Marine Policy*, X (1986), pp. 132–54 at 149.

66 This point was made by Traavik and Østreng as long ago as 1977, but still appears to be valid, although it may be that the Soviet Union's petroleum needs will not make it valid for so very much longer. See K. Traavik and W. Østreng, 'Security and Ocean Law: Norway and the Soviet Union in the Barents Sea', *Ocean Development and International Law*, IV (1977), pp. 343–6 at 358. Norway appears to be just as keen on an agreement as when Traavik and Østreng wrote, since it takes the view that an agreement would lead to greater stability in the region – something which has been the policy of successive Norwegian governments.

67 D.M. Johnston, *The Theory and History of Ocean Boundary-making*, Kingston and Montreal: McGill-Queen's University Press, 1988, *passim*; D.M. Johnston and P.M. Saunders, *Ocean Boundary Making: Regional Issues and Developments*, London: Croom Helm, 1988, ch. 9; G. Osherenko and O. Young, *The Age of the Arctic*, Cambridge: Cambridge University Press, 1989, pp. 264–5; R.R. Churchill, 'Maritime Delimitation in the Jan Mayen Area', *Marine Policy*, IX (1985), pp. 16–38 at 27–8.

68 For an extended discussion of the problems for military activities raised by offshore oil and gas activities, see B. Kibsgaard, 'Militære operasjoner og mulige interessekonflikter i de nordlige havområdene', *Internasjonal Politikk*, 1988, pp. 35–66, esp. 43–50.

69 That installations can in practice be an obstacle to submarines is shown by the

fact that a West German submarine collided with a Norwegian oil platform in the North Sea in March 1988. This accident was the first of its kind.

70 Tresselt, op. cit., p. 81.

71 For example W. Østreng, 'Delimitation Arrangements in Arctic Seas', *Marine Policy*, X (1986), pp. 132–54 at 153–4; *idem*, 'Soviet–Norwegian Relations in the Arctic', *International Journal*, XXXIX (1984), pp. 866–87 at 884–7.

72 [1969] ICJ Rep., p. 47.

73 *Tacna Arica* arbitration, *American Journal of International Law*, XIX (1925), p. 393 ff. at 398; *Railway Traffic between Lithuania and Poland* case, PCIJ, Ser. A/B, 42 (1931), pp. 108, 116; *Lac Lanoux* arbitration, RIAA, XII (1956), p. 285; *North Sea Continental Shelf* cases, pp. 47–8; *Gulf of Maine* case, p. 292.

74 M. Miyoshi, 'The Basic Concept of Joint Development of Hydrocarbon Resources on the Continental Shelf', *International Journal of Estuarine and Coastal Law*, III (1988), pp. 1–18 at 10; R. Lagoni, 'Oil and Gas in Continental Shelf Areas which are disputed or where the Delimitation has not been effected', paper given at the British Institute of International and Comparative Law's Conference on Joint Development of Offshore Oil and Gas, 1989.

75 *Aegean Sea Continental Shelf* case (*Request for Interim Measures of Protection*), [1976] ICJ Rep., pp. 3 ff. at 10–11.

76 PCIJ, Ser. A/B, 79 (1939), p. 199.

77 League of Nations Treaty Series, XCIII, p. 343.

78 Lagoni, op. cit.

79 Ibid.

80 *Gulf of Maine* case, [1984] ICJ Rep., pp. 282, 286.

81 Text in R.R. Churchill *et al.* (eds), *New Directions in the Law of the Sea*, Dobbs Ferry, NY: Oceana, 1977, V, p. 284.

82 B. Johnson Theutenberg, *Folkrätt och Säkerhetspolitik*, Stockholm: Norstedts, 1986, pp. 244–7, 251–2.

83 Stortingstidende, 1981–2, p. 2678.

84 R. Lagoni, 'Interim Measure pending Maritime Delimitation Agreements', *American Journal of International Law*, LXXVIII (1984), pp. 345–68 at 366.

85 Scrivener, *Oil, Fish and Strategy*, p. 14.

86 Ibid. ; Tresselt, op. cit., p. 82.

87 D. Donders, A. Jørgensen-Dahl and O. Schram Stokke, *Northern and Arctic Boundary Disputes*, Oslo: Fridtjof Nansens Institutt, 1990, p. 18.

88 One writer has suggested that a basic precondition for agreement on a boundary, which is currently missing, is the necessary trust between the parties: see Madsen, op. cit., p. 21.

89 It is interesting to note, by way of comparison, that the USSR has recently settled two other long-standing maritime boundary disputes by concluding agreements with the states concerned, namely Sweden (in 1988, after nineteen years of negotiations) and the USA (in 1990, after nine years of negotiations: see n. 56 above). In neither of these cases were security considerations a factor.

90 In June 1991 the Norwegian Prime Minister announced that considerable progress had recently been made in the boundary negotiations and that agreement had now been reached on three-quarters of the disputed area, *UD-informasjon*, 20 (1991).

4 FISHERIES MANAGEMENT AND ACCESS TO FISHERY RESOURCES

1 A.H. Hoel, 'Ressursforvaltningen i Nordområdene', in T.E. Grung (ed.), *Norsk Utenrikspolitisk Årbok 1989*, Oslo: Norsk Utenrikspolitisk Institutt, 1990, p. 35.

2 *Extract of the Report of the Advisory Committee on Fishery Management to the North-east Atlantic Fisheries Commission*, ACFM October–November Meeting, FC 8/1 (pt 2), ICES, 1989.

3 Convention text in S.H. Lay, R. Churchill and M. Nordquist, *New Directions in the Law of the Sea*, and I, Dobbs Ferry, New York: Oceana Publications, Inc., 1973, p. 395. On fisheries management under the NEAFC regime, see A. Underdal, *The Politics of International Fisheries Management: The Case of the North-east Atlantic*, Oslo: Universitetsforlaget, 1980.

4 Text in Lay *et al.*, op. cit., p. 484.

5 G. Sætersdal, '200 Mile Zones: Have the Expectations been Fulfilled?', in G. Ulfstein, P. Andersen and R. Churchill, *The Regulation of Fisheries: Legal, Economic and Social Aspects*, Strasbourg: Council of Europe, 1987, pp. 6–21 at 6, with further references.

6 See G. Hardin, 'The Tragedy of the Commons', *Science*, CLXII (1968), pp. 1243–8.

7 *Om lov om Norges økonomiske sone*, Norwegian Government Paper Odelstingsproposisjon 4, 1976–7, p. 2.

8 A.P. Allison, 'The Soviet Union and UNCLOS III: Pragmatism and Policy Evolution', *Ocean Development and International Law*, XVI (1986), pp. 109–36 at 121.

9 Ibid., p. 122; B.H. Oxman, *From Cooperation to Conflict: The Soviet Union and the United States at the Third U.N. Conference on the Law of the Sea*, The McKernan Lectures, Institute for Marine Studies, University of Washington, 1985, p. 11.

10 Text of the Norwegian–United Kingdom Exchange of Notes, 30 January 1975 in R. Churchill, M. Nordquist and S.H. Lay, *New Directions in the Law of the Sea*, V, Dobbs Ferry, New York: Oceana Publications, Inc., 1977, p. 26. The text of the treaties between Norway and the other states referred to is very similar to that of the Norwegian–United Kingdom Exchange of Notes.

11 Ibid., p. 21, for text of the Norwegian Law on Trawler-free zones beyond the Norwegian Fishery Limit of 17 January 1975 and accompanying Regulations of 31 January 1975.

12 B. Fløistad, *Fisk og Utenrikspolitikk, Norges Fiskerisamarbeid med Andre Land i Barentshavet, Norskehavet og Nordsjøen*, Oslo: Fridtjof Nansen Institute, 1987, pp. 46–7. An English version of this paper has been published as Occasional Paper 37, the Law of the Sea Institute, Hawaii, 1989. For full details see the bibliography.

13 United Nations Treaty Series, CMLXXXIII, p. 3.

14 Norwegian Government Paper Sportingsproposisjon 86, 1974–5, p. 1.

15 Fløistad, op. cit., p. 47.

16 B. Fløistad and O. Schram Stokke, 'Common Concerns: National Interests. Norway, the Soviet Union and the Barents Sea Fisheries', *International Challenges*, IX/2 (1989), pp. 23–30 at 25.

17 According to Allison, op. cit., p. 123, the USSR reacted with 'surprise and

anger', whereas Oxman, op. cit., p. 15, claims that 'the Soviets seemed shocked at and betrayed by the U.S. 200-mile zone'.

18 Agreement on Mutual Fishery Relations. English text in Churchill *et al.*, op. cit., p. 348. The 1975 and the 1976 treaties are discussed in more detail in G. Ulfstein, *Økonomiske soner: hva nå?*, Tromsø: Universitetsforlaget, 1982.

19 Allison, op. cit., p. 123. Text in UN Legislative Series B/19, p. 253 and in W.E. Butler, *The USSR, Eastern Europe and the Development of the Law of the Sea*, Dobbs Ferry, New York: Oceana Publications, Inc., booklet F1, 1983.

20 Text in M. Nordquist, S.H. Lay and K.R. Simmonds, *New Directions in the Law of the Sea*, VII, Dobbs Ferry, New York: Oceana Publications, Inc., 1980, p. 199.

21 E. Franckx, 'New Soviet Fishery Regulations concerning the EEZ: An Appraisal', *Marine Policy*, XI (1987), pp. 125–32 at 127.

22 Text in Churchill *et al.*, op. cit., p. 337.

23 Fløistad and Schram Stokke, op. cit., p. 25.

24 R.R. Churchill and A.V. Lowe, *The Law of the Sea*, Manchester: Manchester University Press, 2nd edn, 1988, p. 146.

25 Sætersdal, op. cit., p. 11.

26 This was earlier commonly assumed to be the case. However, in 1991 French and Greenlandic trawlers began fishing in this area (though later in the year they discontinued their fishing). This led to negotiations between Norway and Greenland, which for the time being seem to have solved the problem. This incident demonstrates that this area may in the future cause management problems. On the regulation of fisheries in such areas, see W.T. Burke, 'Fishing in the Bering Sea Donut: Straddling Stocks and the New International Law of Fisheries', *Ecology Law Quarterly*, XVI (1989), pp. 285–310.

27 Text of the revised NEAFC convention of 1980 in *Official Journal of the European Communities*, L227, 1981, p. 22.

28 Agreement between the Government of Norway and the Government of the Union of Soviet Socialist Republics on Measures for Regulating the Catch and Conserving Stocks of Seals in the North-eastern Part of the Atlantic Ocean, Oslo: 22 November 1957, see United Nations Treaty Series, CCCIX (1958), p. 280.

29 Cf. M. Dahmani, *The Fisheries Regime of the Exclusive Economic Zone*, Dordrecht: Martinus Nijhoff Publishers, 1987, pp. 114–15.

30 See G. Sætersdal, *Principles of Determining Allowable Catches with Particular Reference to Shared Resources*, FAO Fisheries Report, 206, 1978, suppl. 1.

31 Norwegian Government Paper Stortingsproposisjon 38, 1973–4; also in R. Churchill and M. Nordquist, *New Directions in the Law of the Sea*, IV, Dobbs Ferry, New York: Oceana Publications, Inc., 1975, p. 175.

32 J. Hamre, *Fiskebestandene i Norske Farvann: Naturgrunnlag og Beskatning*, Tromsø: Norwegian School of Fisheries, 1986, p. 112.

33 Hoel, op. cit., p. 36.

34 Churchill and Lowe, op. cit., p. 233.

35 Cf. W.T. Burke, 'The Law of the Sea Convention and Fishing Practices of Nonsignatories, with Special Reference to the United States', in J.M. Van Dyke (ed.), *Consensus and Confrontation: The United States and the Law of*

the Sea Convention, Honolulu, Hawaii: The Law of the Sea Institute, University of Hawaii, 1985, pp. 314–37 at 319: 'Thus, the coastal state is authorized to decide that 'optimum utilization' means fishing at whatever level is compatible with the allowable catch; it is not the responsibility of the coastal state to fix the allowable catch so that a particular level of fishery harvest is secured.'

36 Churchill and Lowe, op. cit., p. 233.

37 G. Ulfstein, '200-Mile Zones and Fisheries Management', *Nordisk Tidsskrift for International Ret*, LII/3–4 (1983), pp. 3–33 at 16–17.

38 *Norsk Polarforskning*, Norwegian Government Publications, Nov. 1989: 9, p. 18.

39 Hoel, op. cit., p. 35.

40 Hamre, op. cit., p. 64.

41 'Ressursoversikten 1989', *Fiskets Gang*, 3 (1989), pp. 10–11.

42 Ibid., p. 13.

43 *Extract of the Report of the Advisory Committee on Fishery Management to the North-east Atlantic Fisheries Commission*, ACFM October–November Meeting, 1989, International Council for Exploration of the Sea, FC 8/1 (pt 2), p. 3.

44 G. Sætersdal, 'Fish Resources Research and Fishing Management: A Global Perspective', paper presented at the ICES Statutory Meeting, Bergen, October 1988, p. 2.

45 O. Nakken, 'Bestandssituasjon og Forvaltningshistorikk for Fisk og Reker i Barentshavet og Tilstøytande Område', in *Barentshavet: Vårt Felles Skattkammer*, Conference Report, Honningsvåg, 23–4 August 1989, p. 50. See also *Norsk Polarforskning*, p. 18.

46 T. Jakobsen, 'Ressurssituasjonen før, nå og framtida i torskefiskeriene', *Me'a*, 7–8(1990), p. 14.

47 J. Hamre, 'Barentshavets økologi følsom overfor Beskatning', *Fiskets Gang*, 9 (1989), p. 16.

48 G. Sætersdal, 'Fish Resources Research', p. 8.

49 S. Tjelmeland, 'Fleirbestandsforvaltning i eit økologisk Perspektiv', in *Barentshavet: Vårt Felles Skattkammer*, p. 131.

50 Nakken, op. cit., p. 47.

51 G. Munro, 'The Management of Shared Fishery Resources under Extended Jurisdiction', in Ulfstein *et al.*, op. cit., pp. 27–45 at 34.

52 *Norsk Retstidende* 1983, p. 1213; 1984, p. 1379.

53 Forskrift om endring av forskrift om fiske i fiskevernsonen ved Svalbard, 8 August 1989, *Norsk Lovtidend*, 1989, p. 749.

54 Nakken, op. cit., p. 49.

55 P.J. Derham, 'The Implementation and Enforcement of Fisheries Legislation', in Ulfstein *et al.*, op. cit., pp. 71–81 at 71.

56 Ibid., p. 81.

57 The Norwegian newspaper *Aftenposten*, 17 October 1989.

58 The Norwegian newspaper *Nordlys*, 18 October 1989.

59 See W.T. Burke, '1982 Convention on the Law of the Sea Provisions on Conditions of Access to Fisheries Subject to National Jurisdiction', FAO Fisheries Report 293, 1983, pp. 23–42. See also S. Garcia, J.A. Gulland and E. Miles, 'The New Law of the Sea, and the Access to Surplus Fish Resources: Bioeconomic Reality and Scientific Collaboration', *Marine Policy*, X (1986), pp. 192–201 and Dahmani, op. cit., p. 55.

60 Dahmani, op. cit., p. 67.
61 Churchill and Lowe, op. cit., pp. 233, 321.
62 Fløistad, op. cit., p. 36.
63 Agreement on an Interim Practical Arrangement for Fishing in an Adjoining Area in the Barents Sea, *Overenskomster med fremmede stater*, 1978, p. 436.
64 Fløistad, op. cit., p. 52.
65 Norwegian Government Paper, Stortingsproposisjon 70 (1977–8), *Avtale mellom Norge og Sovjetunionen om en midlertidig praktisk ordning for fisket i et tilstøtende område i Barentshavet*, p. 3.
66 Fløistad, op. cit., p. 55.
67 It is, however, incorrect to claim that the Agreement does not apply to fishing of Atlanto-Scandic herring, as claimed by E. Hey, *The Regime for the Exploitation of Transboundary Marine Fisheries Resources*, Dordrecht: Martinus Nijhoff Publishers, 1989, p. 170. This author explains, however, the more detailed provisions on such fishing correctly.
68 Hey, op. cit., p. 170 claims incorrectly that 'The agreement does not contain any provisions relating to enforcement and control activities in relation to vessels flying the flag of third states'.
69 S. Andresen, *Makt og Rett på Hav og Havbunn*, Oslo: Universitetsforlaget, 1987, p. 120.
70 *Norsk Lovtidend*, 1977, p. 508.
71 The Norwegian Foreign Minister, Knut Frydenlund, in Norwegian Parliamentary Records, Stortings forh. 1976–7, p. 4327.
72 Fløistad, op. cit., p. 64.
73 *Norsk Lovtidend*, 1978, p. 275.
74 *Norsk Lovtidend*, 1985, p. 1377.
75 *Norsk Lovtidend*, 1982, p. 865.
76 The quota regulations are discussed in G. Ulfstein, 'Reguleringsproblemer i fiskevernsonen ved Svalbard', *Lov og Rett* (1987), pp. 336–44.
77 F. Bergesen, 'Fiskevernsonen i praksis', in G. Ulfstein (ed.), *Fiskevernsonen ved Svalbard*, Tromsø: Institutt for fiskerifag, Universitetet i Tromsø, 1983, pp. 99–115 at 105.
78 Forskrift om regulering av torskefiske i fiskevernsonen ved Svalbard i 1990, *Norsk Lovtidend*, 1990, p. 432.
79 For further information about these incidents, see Fløistad, op. cit., p. 124–7.
80 The Norwegian newspaper *Nordlys*, 25 August 1988, 27 August 1988 and 5 January 1989.
81 *UD-informasjon*, 17, 2 June 1989, p. 35.
82 The Norwegian newspaper *Fiskeribladet*, 14 September 1989.
83 Hoel, op. cit., p. 38.
84 *UD-informasjon*, 16, 1987.
85 *Norsk Lovtidend*, 1989, p. 748.
86 The Norwegian Foreign Minister, Mr Stoltenberg, stated in October 1987 that the situation in the Svalbard zone 'throws shadows ahead, when the time comes for exploitation of the continental shelf resources', *UD-informasjon*, 21, p. 5.
87 *Norsk Polarforskning*, p. 22.
88 R. Barrett and T. Haug, 'Fugl og pattedyr: topp-predatorer i Barentshav-systemet', *Ottar*, 1 (1989), p. 32.

89 Ibid., pp. 32–5.
90 Hoel, op. cit., p. 40.
91 The IWC was established by the International Convention for the Regulation of Whaling, Washington, 2 December 1946. Text in United Nations Treaty Series, CLXI, p. 72; Lay *et al.*, op. cit., p. 418.
92 S. Andresen, 'Science and Politics in the International Management of Whales', *Marine Policy*, XIII (1989), pp. 99–118 at 111.
93 Ibid.
94 J. Mønnesland *et al.*, *Whaling in Norwegian Waters in the 1980s: The Economic and Social Aspects of the Whaling Industry, and the Effects of its Termination*, Oslo: Norwegian Institute for Urban and Regional Research, 1990, p. 8.
95 O. Flaaten, *The Economics of Multispecies Harvesting: Theory and Application to the Barents Sea Fisheries*, Berlin: Springer-Verlag, 1989, p. 114.
96 Mønnesland *et al.*, op. cit., pp. 3, 38–9.
97 Andresen, op. cit., p. 114.
98 Ibid., pp. 111–12. An exchange of letters between the USA and Norway on the Norwegian catch of whales for research purposes took place in 1990, *inter alia* to avoid export restrictions: see Norwegian Ministry of Foreign Affairs, Press Release 22, 1990, 16 February 1990.
99 Barrett and Haug, op. cit., pp. 37–9.
100 *Forskningsprogram for sjøpattedyr*, Norwegian Fisheries Research Council, Trondheim, 1988, p. 2.
101 Hoel, op. cit., p. 40.
102 Text in British Government Paper *Cmnd.*, 8830, 1983.

5 OFFSHORE PETROLEUM ACTIVITIES

1 Norwegian Government Paper Stortingsmelding 46 (1986–7), p. 5.
2 Ibid.
3 See, for example, H.O. Bergesen, A. Moe and W. Østreng, *Soviet Oil and Security Interests in the Barents Sea*, London: Frances Pinter, 1987, esp. chs 1 and 4; and S. Lewarne, *Soviet Oil: The Move Offshore*, Boulder, Colo.: Westview Press, 1988, esp. ch. 1. The analysis that follows is based on these works.
4 Lewarne, op. cit., p. 30.
5 Bergesen *et al.*, pp. 33–5, 49–51; *Oil and Gas Journal*, LXXXVI/37 (1988), p. 32.
6 *Oil and Gas Journal*, LXXXVI 19 (1988), p. 16.
7 For a fuller analysis of the possible conflicts between the fishing and offshore petroleum industries, see R.R. Churchill, 'The Conflict between the Fishing and Offshore Petroleum Industries: Experience in the North Sea', *Oil and Gas Law and Taxation Review*, VIII/3 (1990), pp. 62–9; and G. Ulfstein, 'The Conflict between Petroleum Production, Navigation and Fisheries in International Law', *Ocean Development and International Law*, XIX (1988), pp. 229–62.
8 Art. 5(6) of the 1958 Convention; art. 60(7) of the 1982 Convention.
9 See also Ulfstein, op. cit., esp. pp. 232–5.
10 The leading study is E.D. Brown, *Sea-bed Energy and Mineral Resources and the Law of the Sea*, London: Graham and Trotman, 1984, I.

11 Art. 78 of the 1982 UN Convention on the Law of the Sea.
12 Arts 3 and 5(1) of the 1958 Convention on the Continental Shelf; art. 78 of the 1982 Convention.
13 Art. 5(2) (3) and (4) of the 1958 Convention; art. 60(1) (2) (4) (5) and (6) of the 1982 Convention.
14 Art. 5(5) and (6) of the 1958 Convention; art. 60(3) and (7) of the 1982 Convention.
15 IMO Assembly Resolution A.672(16).
16 Art. 82 of the 1982 Convention.
17 The classic study, though now somewhat dated, is C.A. Fleischer, *Petroleumsrett*, Oslo: Universitetsforlaget, 1983. A more recent work is E. Selvig, *Petroleumsrett til Studiebruk*, Oslo: Sjørettsfondet, 1988.The major current piece of primary legislation is the Petroleum Law 1985, which replaces and repeals much earlier legislation.
18 Petroleum Law, s.5.
19 Ibid., s.6.
20 Ibid., s.8.
21 Ibid.
22 Norwegian Government Paper Stortingsmelding 40, 1988–9.
23 *Oil and Gas Journal*, LXXXVIII 18 (1990), p. 18. As an exception to the procedures required under s.7 of the Petroleum Law, a few so-called 'strategic blocks' had from 1986 been opened and allocated in the Barents Sea so as to allow more continuity in exploration for oil companies and to try to get some better idea of prospects in the Barents Sea.
24 The current regulations are to be found in Royal Decree of 14 June 1985. Given the weather, ice and related conditions prevailing in the Barents Sea, it may be necessary to make the present regulations more stringent for offshore oil and gas activities in the Barents Sea.
25 For a fuller study see Churchill, op. cit.
26 Royal Decree of 14 June 1985, s.15.
27 Law No. 35 of 9 June 1989 amending Law No. 11 of 22 March 1985 on Petroleum Activities.
28 Royal Decree of 26 October 1979.
29 UN Legislative Series B/15, p. 443. See also to the same effect the Regulations of 11 January 1974, UN Legislative Series B/18, p. 169.
30 But in 1978 it was decided to concentrate all offshore activities in the Ministry of Gas: see Lewarne, op. cit., pp. 57–60. Whether the 1969 Regulations have been amended to reflect this fact the authors have not been able to ascertain.
31 For further details of Soviet offshore administration, see Bergesen *et al.*, op. cit., pp. 41–5, 106–25.
32 Para. 5. Text in UN Legislative Series B/15, p. 441.
33 Para. 2. Text in UN Legislative Series B/16, p. 167.
34 W.E. Butler, 'Joint Ventures and the Soviet Arctic', *Marine Policy*, XIV (1990), pp. 169–76 at 175.
35 According to Scrivener, the 1987 legislation was not very attractive to Western companies generally because of its provisions on share ownership and taxation of profits in hard currency repatriated abroad. Some of these impediments were removed by the 1989 legislation. In the context of the Barents Sea specifically, security considerations may lead the USSR not to seek use of the opportunities presented by the legislation, and economic

considerations may deter Western companies from doing so. See D. Scrivener, *Gorbachev's Murmansk Speech: The Soviet Initiative and Western Response*, Oslo: The Norwegian Atlantic Committee (DNAK), 1989, pp. 20–2, 38–40.

36 *Oil and Gas Journal*, XXXVIII/32 (1990), p. 28.

37 One limitation on such activities by Western companies is the so-called COCOM rules, which restrict exports of strategic technology to the USSR.

38 See, for example, Bergesen *et al.*, op. cit., p. 102; Lewarne, op. cit., pp. 143–54; and Scrivener, op. cit., p. 21.

39 Resolution No. 564 of 18 July 1969, UN Legislative Series B/15, p. 443, paras 4 and 6.

40 See, for example, the 'Nature' programme on BBC2 on 3 December 1990.

41 This matter is the subject of a recently completed full-length study: D. Brubaker, *International Management of Pollution in the Barents Sea*, School of Law, University of Tromsø, 1991.

42 Paris Commission, *Tenth Annual Report*, London, 1989, p. 199. The average figure given is based on the figures for 1984–6 inclusive.

43 Lewarne, op. cit., p. 108.

44 Production water is salt water which is always found to be present with oil when it reaches the platform. In addition, water is pumped into an oil well as it is exploited in order to maintain pressure in the well. This water eventually forms part of the production water.

45 Convention for the Prevention of Marine Pollution from Landbased Sources, 1974. Text in UK Treaty Series 64, 1974.

46 Parcom Decision 86/2 of 4 June 1986 on the Use of Oil-based Muds. Text in Paris Commission, *Eighth Annual Report*, London, 1986, p. 58.

47 Parcom Decision 88/1 of 17 June 1988 on the Use of Oil-based Muds. Text in Paris Commission, *Tenth Annual Report*, London, 1989, p. 95.

48 Here, and in the discussion that follows, where the authors say that they have not been able to ascertain particular information relating to Soviet law and practice, it is also the case that Brubaker, op. cit., has been unable to discover the information in question, even though – as the authors are aware – he went to considerable efforts to do so.

49 Reproduced in W.E. Butler, *The USSR, Eastern Europe and the Development of the Law of the Sea*, Dobbs Ferry, NY: Oceana Publications, booklet J4, 1986.

50 Paris Commission, *Eighth Annual Report*, pp. 11–12.

51 Text in R. Churchill *et al.* (eds), *New Directions in the Law of the Sea*, Dobbs Ferry, NY: Oceana Publications, IV, 1975, p. 345.

52 UN Legislative Series B/15, p. 443.

53 Butler, op. cit., booklet J4.

54 Churchill *et al.*, op. cit., p. 345.

55 Convention on the Prevention of Marine Pollution by Dumping of Waste and Other Matter. Text in Churchill *et al.*, op. cit., p. 331.

56 Royal Decree of 26 October 1979.

57 Butler, op. cit., booklet F3, 1986.

58 It may also be noted that the 1984 Edict on Intensifying Nature Protection in Areas of the Far North and Marine Areas Adjacent to the Northern Coast of the USSR (see Butler, op. cit.) calls for the establishment of special rules for the design, construction and operation of offshore installations in the marine areas adjacent to the northern coast of the USSR.

59 J.S. Gray, *The Implementation of the North Sea Declarations. An Assessment of the Norwegian Follow-up*, Greenpeace Paper 13, Implementation of the Second North Sea Conference, 1990, pp. 25–8.

60 Not apparently yet published. Text obtained from the Norwegian Ministry of the Environment.

61 Brubaker, op. cit., p. 158. For details of co-operation hitherto, see *Aftenposten*, 1 and 23 February 1990.

62 Canada–Denmark. Agreement for Co-operation relating to the Marine Environment, 1983. *International Legal Materials*, XXIII (1984), p. 269.

63 Canada–USA. Agreement relating to the Addition of Annex IV concerning the Beaufort Sea to the Joint Marine Pollution Contingency Plan promulgated pursuant to the Agreement of 19 June 1974, 1977. Treaties and other International Acts Series 8957.

64 USSR–USA. Agreement concerning Co-operation in Combating Pollution in the Bering and Chukchi Seas, 1989.

65 Such a bilateral treaty could, of course, be widened to encompass all the pollution matters discussed above. This would be especially desirable if the USSR does not become a party to the Paris Convention.

66 Perhaps along the lines of the arrangements for the North Sea contained in the Agreement for Co-operation in Dealing with Pollution of the North Sea by Oil and Other Harmful Substances, 1983 British Government Paper, Cmnd. 9104.

6 CONCLUSIONS

1 J. Sanness, 'Rett, makt og fornuft i nordpolitikken', *Internasjonal Politikk* (1978), 3B, pp. 447–59 at 457.

2 G. Ulfstein, 'Sikkerhetspolitikk og næringsinteresser', *Internasjonal Politikk* (1979), 2, pp. 327–30 at 328.

3 C.A. Fleischer, *Petroleumsrett*, Oslo: Universitetsforlaget, 1983, p. 223.

4 Ibid.

5 We are therefore not convinced that the USSR is as interested as Norway in the need for unlimited application of Norwegian licensing conditions on the Svalbard shelf, as claimed in W. Østreng, *Det Politiske Svalbard*, Oslo: Gyldendal, 1975, pp. 91–2.

6 See the late Norwegian Foreign Minister K. Frydenlund, *Lille land – hva nå?*, Oslo: Universitetsforlaget, 1982, p. 98.

7 W. Østreng, 'Delimitation arrangements in Arctic seas', *Marine Policy*, X (1986), pp. 132–54 at 153.

8 Ibid.

Bibliography

Note: this bibliography includes only items which are specifically, though not necessarily exclusively, concerned with the Barents Sea.

Andenæs, J., *Suverenitet og Eiendomsrett på Svalbard*, Oslo: Universitetsforlaget, 1984.

Archer, C., 'Russia's Arctic dimension', *The World Today*, XLIV (1988), pp. 47–50.

—— (ed.), *The Soviet Union and Northern Waters*, London: Routledge, 1988.

Archer, C. and Scrivener, D., 'Frozen frontiers and resource wrangles: conflict and co-operation in Northern Waters', *International Affairs*, LIX (1983), pp. 59–76.

—— (eds), *Northern Waters: Resources and Security Issues*, London: Croom Helm, 1986.

Armstrong, T., Rogers, G. and Rowley, G., *The Circumpolar North: A Political and Economic Geography of the Arctic and Subarctic*, London: Methuen, 1978.

Barrett, R and Haug, T., 'Fugl og pattedyr: topp-predatorer i Barentshavsystemet', *Ottar*, 1 (1989), p. 32.

Bergesen, H.O., Moe, A. and Østreng, W., *Soviet Oil and Security Interests in the Barents Sea*, London: Pinter, 1987.

Bergsager, E., 'Barentshavet: ressurser og utvinningsmuligheter', *Internasjonal Politikk* (1988), pp. 7–34.

Bernhardt, J.P.A., 'Spitzbergen: jurisdictional friction over unexploited oil reserves', *California Western International Law Journal*, IV (1973), pp. 65–120.

Bomsdorf, F., 'Norway's Nordpolitikk and the Soviet Union', *Aussenpolitikk*, XXX (1979), pp. 258–73.

Børde, H. (ed.), *Svalbard og Havområdene*, Oslo, 1983.

Bretton, P., 'L'Union sovietique et la clause de non militarisation du Spitsberg', *Annuaire français de droit international*, XI (1965), pp. 122–44.

Brigham, L.W. (ed.), *The Soviet Maritime Arctic*, London: Bellhaven Press, 1991.

Bruel, E., 'Orienterende oversigter over aktuelle spørsmaal (Svalbard)', *Nordisk Tidsskrift for Internasjonal Ret* (1947–8), pp. 32–42.

Brundtland, A.O., 'Fisk og sikkerhet: noen trekk ved det norsk–sovjetiske forhold 1979', *Norsk Utenrikspolitisk Årbok* (1979), pp. 67–77.

—— 'Den nye sovjetiske nordpolitikken og mulige norske svar', *Internasjonal Politikk* (1988), pp. 95–134.

Butler, W.E., 'Soviet fishery jurisdiction in the Arctic', *Polar Record*, XVIII (1977), pp. 575–9.

—— *Northeast Arctic Passage*, Alphen aan den Rhijn: Sijthoff & Noordhoff, 1978.

—— 'Joint ventures and the Soviet Arctic', *Marine Policy*, XIV (1990), pp. 169–76.

Churchill, R.R., 'The maritime zones of Spitsbergen', in W.E. Butler (ed.), *The Law of the Sea and International Shipping: Anglo-Soviet Post-UNCLOS Perspectives*, London: Oceana, 1985, pp. 189–234.

—— 'Maritime boundary problems in the Barents Sea', in G. Blake (ed.), *Maritime Boundaries and Ocean Resources*, London: Croom Helm, 1987, pp. 147–161.

Churchill, R.R. and Ørebech, P., 'Om norsk lovgivnings anvendelse på Svalbard', *Lov og Rett* (1985), pp. 164–177.

Conférence de la Paix, (1919–20), *Commission du Spitsberg*, Recueil des actes de la conférence, Pt VII, Paris, 1924.

Continental Shelf Boundary Norway–Soviet Union, Washington DC: US Department of State, International Boundary Study Series A, Limits in the Seas 17, 1970.

Donders, D., Jørgensen-Dahl, A. and Stokke, O.S., *Northern and Arctic Boundary Disputes*, Oslo: Fridtjof Nansen Institute, 1990.

Flaaten, O., *The Economics of Multispecies Harvesting: Theory and Applications to the Barents Sea Fisheries*, Berlin: Springer Verlag, 1989.

Fleischer, C.A., 'Oil and Svalbard', *Nordisk Tidsskrift for Internasjonal Ret*, XLV (1976), pp. 7–13.

—— 'Svalbards folkerettslige stilling', in A. Treholt *et al.* (eds), *Norges havretts-og ressurspolitikk*, Oslo: Tiden Norsk Forlag, 1976, pp. 139–155.

—— 'Le régime d'exploitation de Spitsberg (Svalbard)', *Annuaire français de droit international*, XXIV (1978), pp. 275–300.

—— 'The question of earlier treaty rights applied to new maritime zones: Spitsbergen and the law of the sea', *Iranian Review of International Relations*, (1978), pp. 245–62.

—— 'The Northern Waters and the new maritime zones', *German Yearbook of International Law*, XXII (1979), pp. 100–18.

—— *Petroleumsrett*, Oslo: Universitetsforlaget, 1983.

Fløistad, B., *Fisk og Utenrikspolitikk: Norges Fiskerisamarbeid med Andre Land i Barentshavet, Norskehavet og Nordsjøen*, Oslo: Fridtjof Nansen Institute, 1987. (Eng. trans. as *Fish and Foreign Policy: Norway's Fisheries Policy towards Other Countries in the Barents Sea, the Norwegian Sea, and the North Sea*, Honolulu, Hawaii: Law of the Sea Institute Occasional Paper 37, 1989.)

Fløistad, B. and Stokke, O.S., 'Common concerns – national interests: Norway, the Soviet Union and the Barents Sea fisheries', *International Challenges*, IX (1989), pp. 23–30.

Franckx, E., 'New Soviet fishery regulations concerning the EEZ: an appraisal', *Marine Policy*, XI (1987), pp. 125–32.

Goldstein, W.(ed.), *Clash in the North*, London: Pergamon–Brassey's, 1987.

Grøndahl, J.S., *Aktuelle Spørsmal om Svalbard*, Oslo: DNAK [The Norwegian Atlantic Committee], 1982.

Grove, E. (ed.), *NATO's Defence of the North*, London: Brassey's Atlantic Commentaries 1, 1980.

Hagen, L., 'Problems arising from development of petroleum resources in Norway's Arctic areas', in *Energy Law '86*, New York, 1986, pp. 419–35.

Hamre, J., 'Barentshavets økologi følsom overfor beskatning', *Fiskets Gang*, 9 (1989), p. 16.

Hoel, A.H., 'Ressursforvaltningen i nordområdene', *Norsk Utenrikspolitisk Arbok* (1989), pp. 32–44.

Holst, J.J., 'Norway's search for a Nordpolitik', *Foreign Affairs*, LX (1981), pp. 63–86.

Holst, J.J., Hunt, K. and Sjaastad, A.C. (eds), *Deterrence and Defense in the North*, Oxford: Oxford University Press, 1986.

Horesma, P., *The Northern Frontier: Soviet Polar Policy since 1917 and its Relation to the History of Exploration*, Groningen: Universiteitsdrukkerij, 1988.

Jenisch, U., 'Sovereign rights in the Arctic: Maritime policies and practices after UNCLOS III', *German Yearbook of International Law*, XXVIII (1985), pp. 297–321.

Jervell, S. and Nyblom, K. (eds), *The Military Build-up in the High North: American and Nordic Perspectives*, London: University Press of America, 1987.

Johnson Theutenberg, B., 'The Arctic Law of the Sea', *Nordisk Tidsskrift for Internasjonal Ret*, LII (1983), pp. 3–39.

—— *The Evolution of the Law of the Sea: A Study of Resources and Strategy with Special Regard to the Polar Regions*, Dublin: Tycooly International, 1984.

Jørgensen-Dahl, A., 'The Soviet–Norwegian Maritime Disputes in the Arctic: Law and Politics', *Ocean Development and International Law*, XXI (1990), pp. 411–29.

Kellberg, L., 'The Spitzbergen Treaty of 1920', *Nordisk Tidsskrift for International Ret*, LII (1983), pp. 40–81.

Kibsgaard, B., 'Militære operasjoner og mulige interessekonfliker i de nordlige havområdene', *Internasjonal Politikk* (1988), pp. 35–66.

Kjølaas, H., 'Forhandlingene om den grå sonen', *Norsk Utenrikspolitisk Årbok* (1977), pp. 31–70.

Klepsvik, A.H., 'Norges kontinentalsokkel og eksklusiv økonomiske sone', *Internasjonal Politikk* (1979), pp. 343–92.

Lansing, R., 'A unique international problem', *American Journal of International Law* XI (1917), pp. 763–71.

Laursen, F., 'Security aspects of Danish and Norwegian law of the sea policies', *Ocean Development and International Law* XVIII (1987), pp. 199–233.

Lewarne, S., *Soviet Oil: The Move Offshore*, London: Westview Press, 1988.

Lindsey, G., *Strategic Stability in the Arctic*, Oxford: Brassey, Adelphi Paper 241, 1989.

Mathisen, T., *Svalbard i Internasjonal Politikk 1871– 1925*, Oslo: H. Aschehoug & Co., 1951.

—— *Svalbard in the Changing Arctic*, Oslo: Gyldendal Norsk Forlag, 1954.

Moe, A., *Utenrikspolitiske Betingelser og Norsk Svalbard Politikk*, Oslo: Fridtjof Nansen Institute, Paper R005–1983, 1983.

—— 'Barentshavet i sovjetisk energipoltikk', *Internasjonal Politikk* (1988), pp. 135–54.

Morelli, E., *Il regime giuridico dello Svalbard e il nuovo diritto del mare*, Milan: Dott. A. Giuffre, 1988.

Mottolä, K. (ed.), *The Arctic Challenge: Nordic and Canadian Approaches to Security and Co-operation in an Emerging International Region*, Boulder, Colo.: Westview Press, 1988.

Nakken, O., 'Bestandssituasjon og forvaltningshistorikk for fisk og reker i Barentshavet og tilstøytande omrade', in *Barentshavet: Vårt Felles Skattkammer*, Conference Report, Honningsvåg, Norway, August 1989, p. 50.

Nielsen, F.K., 'The Solution of the Spitzbergen Question', *American Journal of International Law*, XIV (1920), pp. 232–5.

Østreng, W., *Økonomi og Politisk Suverenitet, Interessespillet om Svalbards Politiske Status*, Oslo: Universitetsforlaget, 1974.

—— *Det Politiske Svalbard*, Oslo: Universitetsforlaget, 1975.

—— *Politics in High Latitudes: The Svalbard Archipelago*, London: Hurst, 1977.

—— *Svalbards Folkerettslige og Politiske Status*, Oslo: Fridtjof Nansen Institute, Paper AA: P161, 1977.

—— *The Exercise of Norwegian Authority on Svalbard*, Oslo: Fridtjof Nansen Institute, Paper AA:P181, 1978.

—— 'The continental shelf issues in the 'Eastern' Arctic Ocean: Implications of UNCLOS III, with special reference to the Informal Composite Negotiating Text', in J.K. Gamble (ed.), *Law of the Sea: Neglected Issues*, Honolulu, Hawaii: Law of the Sea Institute 1979, pp. 165–82.

—— 'Norwegen und die Sowjetunion in der Barentssee', *Europa-Archiv*, XXIII (1980), pp. 711–20.

—— *Traktatfortolkning og Nasjonale Interesser: Svalbardtraktaten av 9. Februar 1920*, Oslo: Fridtjof Nansen Institute, Paper N:027, 1982.

—— 'Soviet–Norwegian relations in the Arctic', *International Journal*, XXXIX (1984), pp. 866–87.

—— 'Delimitation arrangements in Arctic seas', *Marine Policy*, X (1986), pp. 132–154.

—— *The Soviet Union in Arctic Waters*, Honolulu, Hawaii: Law of the Sea Institute Occasional Paper 36, 1987.

—— 'Geography and military confrontation in the North', *International Challenges*, IX/4 (1989), pp. 11–15.

Rolston, S.J. and McDorman, T.L., 'Maritime boundary making in the Arctic region', in D.M. Johnston and D.M. Saunders (eds), *Ocean Boundary Making: Regional Issues and Developments* , London: Croom Helm, 1988, pp. 16–73.

Sanness, J., 'Rett, makt og fornuft i nordpolitikken', *Internasjonal Politikk* (1978), pp. 447–58.

Scovazzi, T., 'Antichi e recenti problemi del regime giuridico dello Spitzbergen', *Studi in onore di Cesare Grassetti* III (1980).

Scrivener, D., *Oil, Fish and Strategy: The USSR, Svalbard and the Barents Sea*, Aberdeen: Aberdeen Studies in Defence Economics Number 27, 1987.

—— *Gorbachev's Murmansk Speech: The Soviet Initiative and Western Response*, Oslo: DNAK (The Norwegian Atlantic Committee), 1989.

Shusterich, K.M., 'International jurisdictional issues in the Arctic Ocean', *Ocean Development and International Law*, XIV (1984), pp. 235–72.

Singh, E.C., *The Spitsbergen (Svalbard) Question: United States Foreign Policy, 1907–1935*, Oslo: Universitetsforlaget, 1980.

Skagestad, O.G., *Norsk Polarpolitikk*, Oslo, 1975.
Skogan, J.K., 'Gråsoneordningen og fremtiden', *Internasjonal Politikk* (1978), pp. 459–69.
—— 'Gråsone, sektorlinje og norsk-sovjetiske forhandlinger', *Internasjonal Politikk* (1979), pp. 465–9.
Sollie, F., 'Grenser og linjer i Barentshavet', *Norsk Utenrikspolitisk Årbok* (1976), pp. 44–57.
—— *Barents Sea Question*, Oslo: Fridtjof Nansen Institute, Paper AA:P190, 1977.
—— *The Economic and Strategic Importance of the Norwegian and Barents Seas*, Oslo: Fridtjof Nansen Institute, Paper AA:P173, 1978.
—— *Havrettsutviklingen og Svalbardsonen*, Oslo: Fridtjof Nansen Institute, Paper AA:P176, 1978.
Sollie, F. and Østreng, W., 'Betydningen av Svalbard-traktatens artikkel 9 i relasjon til den militærstrategiske utvikling i Barentshav-Svalbard-området', *Internasjonal Politikk* (1977).
Stokke, O.S. and Hoel, A.H., *A Bridge over Troubled Waters: International Fisheries Management in the Barents Sea*, Oslo: Fridtjof Nansen Institute, Report R:009–1990, 1990.
Svalbard, Norwegian Parliamentary Paper, Stortingsmelding 40, Norwegian Ministry of Justice, 1985–6.
Traavik, K. and Østreng, W., 'Security and ocean law: Norway and the Soviet Union in the Barents Sea,' *Ocean Development and International Law*, IV (1977), pp. 343–67.
Tresselt, P., 'Norsk–sovjetiske forhandlinger om avgrensning av kontinentalsokler og økonomiske soner', *Internasjonal Politikk* (1988), pp. 75–94.
Ulfstein, G., 'Sikkerhetspolitikk og næringsinteresser', *Internasjonal Politikk* (1979), pp. 327–30.
—— (ed.): *Fiskevernsonen ved Svalbard*, Tromsø: Universitetet i Tromsø, 1983.
—— 'Reguleringsproblemer i fiskevernsonen ved Svalbard', *Lov og Rett* (1987), pp. 336–44.
Vedrørende Svalbard, Norwegian Parliamentary Paper, Stortingsmelding 39, Norwegian Ministry of Justice, 1974–5.

Index

Life Without Cheap Sunglasses

To Dorothy,

12/16/00